A Different Universe

A
Different
Universe

(REINVENTING
PHYSICS
from the Bottom Down)

Robert B. Laughlin

BASIC
BOOKS

A Member of the Perseus Books Group
New York

Books published by Basic Books are available at special discounts for bulk
purchases in the United States by corporations, institutions, and other
organizations. For more information, please contact the Special Markets
Department at the Perseus Books Group, 11 Cambridge Center, Cambridge
MA 02142, or call (617) 252–5298 or (800) 255–1514, or e-mail
special.markets@perseusbooks.com.

Designed by Brent Wilcox
Text set in 11.75 point Minion

Library of Congress Cataloging-in-Publication Data
Laughlin, Robert B.
    A different universe : reinventing physics from the bottom down /
Robert B. Laughlin.
        p.   cm.
    Includes bibliographical references and index.
    ISBN 0–465–03828–X (hardcover : alk. paper)
    1. Physics—Popular works.   I. Title.
    QC24.5.L38   2005
    530—dc22

                                                        2004028059

05 06 07 / 10 9 8 7 6 5 4 3 2 1

To
Anita

Not only is the universe stranger than we imagine, it is stranger than we can imagine.

Sir Arthur Eddington

CONTENTS

PREFACE

*All the rivers run into the sea; yet the sea is not full; unto the place
from whence the rivers come, thither they return again.*

Eccl. 1:7

There are two conflicting primal impulses of the human mind—
one to simplify a thing to its essentials, the other to see through
the essentials to the greater implications. All of us live with this
conflict and find ourselves pondering it from time to time. At the
edge of the sea, for example, most of us fall into thoughtfulness
about the majesty of the world even though the sea is, essentially,
a hole filled with water. The vast literature on this subject, some of
it very ancient, often expresses the conflict as moral, or as tension
between the sacred and the profane. Thus viewing the sea as simple and finite, as an engineer might, is animistic and primitive,
whereas viewing it as a source of endless possibility is advanced
and human.

But the conflict is not just a matter of perception: it is also physical. The natural world is regulated both by the essentials and by powerful principles of organization that flow out of them. These
principles are transcendent, in that they would continue to hold even
if the essentials were changed slightly. Our conflicted view of nature
reflects a conflict in nature itself, which consists simultaneously of

The essence of life.

primitive elements and stable, complex organizational structures
that form from them, not unlike the sea itself.

The edge of the sea is also a place to have fun, of course, something it
is good to keep in mind when one is down there by the boardwalk being
deep. The real essence of life is strolling too close to the merry-go-
round and getting clobbered by a yo-yo. Fortunately, we physicists are
fully aware of our own sententious tendencies and go to great lengths to
keep them under control. This attitude was artfully expressed in a letter
my colleague Dan Arovas, a faculty member at the University of California at San Diego, wrote to the humor columnist Dave Barry:

> Dear Dave, I am a passionate fan of yours and read your column every
> day. I would give anything to be able to write like you. I have built a
> tree house in your honor and live in it. Yours, Dan

Dan reports that Dave wrote back:

Dear Dan, Thanks for the fan letter. By the way, do they let you anywhere near nuclear weapons? Best, Dave

A few years ago I had occasion to engage my father-in-law, a retired academician, on the subject of the collective nature of physical
law. We had just finished playing bridge late one afternoon and were
working on a couple of gin and tonics in order to escape discussing
movies of emotional depth with our wives. My argument was that reliable cause-and-effect relationships in the natural world have something to tell us about ourselves, in that they owe this reliability to
principles of organization rather than microscopic rules. The laws of
nature that we care about, in other words, emerge through collective
self-organization and really do not require knowledge of their component parts to be comprehended and exploited. After listening carefully, my father-in-law declared that he did not understand. He had
always thought laws cause organization, not the other way around.
He was not even sure the reverse made sense. I then asked him
whether legislatures and corporate boards made laws or were made
by laws, and he immediately saw the problem. He pondered it for a
while, and then confessed that he was now deeply confused about
why things happen and needed to think more about it. Exactly so.

It is a terrible thing that science has grown so distant from the rest
of our intellectual life, for it did not start out that way.[1] The writings
of Aristotle, for example, despite their notorious inaccuracies, are
beautifully clear, purposeful, and accessible.[2] So is Darwin's *Origin of
Species*.[3] The opacity of modern science is an unfortunate side effect
of professionalism, and something for which we scientists are often
pilloried—and deservedly so. Everyone gets wicked pleasure from
snapping on the radio on the drive home from work to hear Doctor
Science give ludicrous answers to phone-in questions such as why
cows stand in the same direction while grazing (they must face Wis-
consin several times a day) and then finish up with, "And remember,

I know more than you. I have a master's degree in science."[4] On another occasion my father-in-law remarked that economics had been terrific until they made it into a science. He had a point.

The conversation about physical law started me thinking about what science had to say about the obviously very unscientific chicken-and-egg problem of laws, organizations of laws, and laws from organization. I began to appreciate that many people had strong views on this subject but could not articulate why they held them. The matter came to a head recently when I realized I was having the same conversation over and over again with colleagues about Brian Greene's *The Elegant Universe*, a popular book describing some speculative ideas about the quantum mechanics of space.[5] The conversation focused on the question of whether physics was a logical creation of the mind or a synthesis built on observation. The impetus for the discussion was never an existential problem, of course, but money, the lack of which is the universal common denominator of world science. But the subject always seemed to drift from there to the pointlessness of making models of the world that were beautiful but predicted no experiments, and from there to the question of what science is. After this happened a number of times in such disparate venues as Seattle, Taipei, and Helsinki, it struck me that the disagreement spawned by Greene's book was fundamentally the same problem that had occupied us that day after bridge. Moreover, it was an ideological dispute: it had nothing to do with what was true and everything to do with what "true" was.

It is commonly said in physics that good notation advances while bad notation retards. This is certainly true. A phonetic alphabet takes less time to master than a pictorial one and thus makes writing more accessible. Decimal numbers are easier to use than roman numerals. The same idea applies to ideologies. Seeing our understanding of nature as a mathematical construction has fundamentally different implications from seeing it as an empirical synthesis. One view identifies us as masters of the universe; the other identifies the universe as the master of us. Little wonder that my colleagues down in the trenches of experimental science had become so animated over this question. At its core the matter is not scientific at all but concerns one's sense of self and place in the world.

The threads of these two world views run very deep. When I was a kid I drove with my parents to Yosemite for a rendezvous with my aunt and uncle, who had driven in from Chicago. My uncle was a brilliant and highly successful patent attorney who seemed to know everything and was not shy about sharing this fact. For example, he once gave me a long sermon on how lasers work after learning that I had just had a lecture on the subject from Charles Townes, the laser's inventor. Evidently, he knew more about it than Professor Townes. On this occasion he and my aunt checked in at the Ahwahnee, the fanciest hotel in the place, held court there with us, consumed a few buffet breakfasts, and then left to drive over Tuolumne Pass to the desert and home. I don't think they saw a single waterfall up close. There was no point, since they had seen waterfalls before and understood the concept. After they left, my family and I hiked up the Merced river, amid the violence and roar, to Nevada Falls and had a picnic on a massive piece of granite next to a meadow full of wildflowers. We understood the concept too but were wise enough not to take our understanding too seriously.

The world view motivating my uncle's attitude toward Yosemite, and arguably also Brian Greene's attitude toward physics, is expressed with great clarity in John Horgan's *The End of Science*, in which he argues that all fundamental things are now known and there is nothing left for us to do but fill in details.[6] This pushes my experimental colleagues beyond their already strained limits of patience, for it is both wrong and completely below the belt. The search for new things always looks like a lost cause until one makes a discovery. If it were obvious what was there, one would not have to look for it!

Unfortunately, this view is widely held. I once had a conversation with the late David Schramm, the famous cosmologist at the University of Chicago, about galactic jets. These are thin pencils of plasma that beam out of some galactic cores to fabulous distances, sometimes several galactic radii, powered somehow by mechanical rotation in the core. How they can remain thin over such stupendous distances is not understood, and something I find tremendously interesting. But David dismissed the whole effect as "weather." He was interested only in the early universe and astrophysical observations that could shed light on it, even if only marginally. He categorized the jets as annoying distractions on the grounds that they had nothing in particular to tell him about what was fundamental. I, in contrast, am fascinated by weather and believe that people claiming not to be are fibbing.

I think primitive organizational phenomena such as weather have something of lasting importance to tell us about more complex ones, including ourselves: their primitiveness enables us to demonstrate with certainty that they are ruled by microscopic laws but also, paradoxically, that some of their more sophisticated aspects are insensitive to details of these laws. In other words, we are able to *prove* in these simple cases that the organization can acquire meaning and life of its own and begin to transcend the parts from which it is made. What physical science thus has to tell us is that the whole being more than the sum of its parts is not merely a concept but a physical phenomenon. Nature is regulated not only by a microscopic rule base but by powerful and general principles of organization. Some of these principles are known, but the vast majority are not. New ones are being discovered all the time. At higher levels of sophistication the cause-and-effect relationships are harder to document, but there is no evidence that the hierarchical descent of law found in the primitive world is superseded by anything else. Thus if a simple physical phenomenon can become effectively independent of the more fundamental laws from which it descends, so can we. I am carbon, but I need not have been. I have a meaning transcending the atoms from which I am made.

The essential elements of this message are articulated in the extensive writings of Ilya Prigogine[7] and even more originally in a famous essay by P. W. Anderson entitled "More is Different"[8] published over 30 years ago. This essay is just as fresh and inspiring today as it was then, and still required reading for any student wishing to work with me.

My views are considerably more radical than those of either of my predecessors, however, because they have been sharpened by recent events. I am increasingly persuaded that *all* physical law we know about has collective origins, not just some of it. In other words, the distinction between fundamental laws and the laws descending from them is a myth, as is the idea of mastery of the universe through mathematics alone. Physical law cannot generally be anticipated by pure thought, but must be discovered experimentally, because control of nature is achieved only when nature allows this through a principle of organization. One might subtitle this thesis the end of reductionism (the belief that things will necessarily be clarified when they are divided into smaller and smaller component parts), but that would not be quite accurate. All physicists are reductionists at heart, myself included. I do not wish to impugn reductionism so much as establish its proper place in the grand scheme of things.

To defend my assertion I must openly discuss some shocking ideas: the vacuum of space-time as "matter," the possibility that relativity is not fundamental, the collective nature of computability, epistemological barriers to theoretical knowledge, similar barriers to experimental falsification, and the mythological nature of important parts of modern theoretical physics. The radicality is, of course, partly a stage prop, for science, as an experimental undertaking, cannot be radical or conservative but only faithful to the facts. But these larger conceptual issues, which are not science at all but philosophy, are often what most interest us because they are what we call upon to weigh merit, write laws, and make choices in our lives.

The objective, then, is not to make controversy for the sake of itself but to help us see clearly what science has become. To do this we must forcibly separate science's function as the facilitator of technology from its function as a means of understanding things—including ourselves. The world we actually inhabit, as opposed to the happy idealization of modern scientific mythology, is filled with wonderful and important things we have not yet seen because we have not looked, or have not been able to look at due to technical limitations. The great power of science is its ability, through brutal objectivity, to reveal to us truth we did not anticipate. In this it continues to be invaluable, and one of the greatest of human creations.

ACKNOWLEDGMENTS

This book would not have come into being without the invaluable efforts of Mr. Steve Lew, whose original concept it was and who worked tirelessly to promote the project with publishers and to encourage me to write. The latter was central, for we scientists have responsibilities and contractual obligations that must be sacrificed to accomplish a task of this size. My interaction with Steve has been one of the most memorable of my long academic career, and I gratefully acknowledge both his extraordinary gifts as a facilitator and organizer and the immense help he provided me in seeing the problem of physical emergence from a humanistic perspective. I also acknowledge his ideas. The tone, form, and scope of the project are partly his, as they emerged out of a series of conversations we had in my office over the course of many months. For all of this, and for helping me edit the manuscript, Steve has my most heartfelt thanks.

I am also deeply indebted to Professor David Pines for much patient help in getting the project off the ground and for critical reading of the manuscript. During David's visit to Stanford in the spring of 1999 we discovered that our views on the physics of collective organization were identical—a great surprise, considering the differences in our backgrounds—as were our perceptions for the need to translate what was so obvious to us into accessible everyday language. This culminated in our coauthoring the essay "The Theory of Everything," in which the main themes of what would eventually

become this book were first articulated.[1] The immense popularity of
that essay, which caught us both off guard, caused us to realize that a
bigger version had to be written. David's visit also induced me to become actively involved in his Institute for Complex Adaptive Matter,
a cross-disciplinary forum dedicated to the worldview that mathematics grows out of experimental observation, not the other way
around. Among its other functions, the institute encourages (forces)
scientists to explain their work to each other in pedestrian terms. The
value of this practice is impossible to overstate. I have learned more
about science through workshops sponsored by this institute and the
personal contacts they generated than I have from all other professional activities combined.

I wish to express special thanks to two institutions that shielded
me from academic duties while I was writing. One is the Institute for
Materials Research in Sendai, Japan, where I spent part of my sabbat-
ical leave in November 2002. I gratefully acknowledge the warm hos-
pitality of Professor Sadamichi Maekawa, with its many fine evenings
of expensive sushi and eel near the banks of the Hirose River. The
other is the Korea Institute for Advanced Study in Seoul, where I am
currently an adjunct professor. My visit there in September 2003 was
especially productive, and I owe my host, Professor C. W. Kim, an
immense debt of gratitude for this—not to mention for the dazzling
variety of restaurants we sampled.

Finally, of course, I must thank my wife, Anita, for her seemingly
endless patience, and the promise that I will indeed now take a break
so that we can make that trip to Maine she has been anticipating for
so long to revisit family haunts and track down some good lobster.

Frontier Law

Nature is a collective idea, and though its essence exist in each indi-vidual of the species, can never its perfection inhabit a single object.

Henri Fuseli

MANY YEARS AGO, WHEN I WAS LIVING NEAR NEW YORK, I attended a retrospective of Ansel Adams, the great nature photographer, at the Museum of Modern Art. Like many people born in the American West, I had always liked Mr. Adams's work and felt I appreciated it better than New Yorkers ever could, so I jumped at the chance to see it firsthand. It was well worth the effort. Anyone seeing these images close up realizes at once that they are not simply sterile pictures of rocks and trees but thoughtful comments on the meaning of things, the immense age of the earth, and the impermanence of human concerns. This exhibition made a much stronger impression on me than I had expected, and it flashes into my mind even now when I am wrestling with a tough problem or having difficulty separating what is important from what is not.

Public television viewers were reminded recently by Ric Burns's excellent *American Experience* documentary that Mr. Adams's work, like any other art, was as much a creation of a specific time and place as of the artist himself.[1] In the early part of the twentieth century,

In Europe, the myth of the frontier is often dismissed as quaint provincialism.

when Adams was a boy and the frontier had been declared closed, Americans debated vigorously over what its loss implied for their future.[2] In the end, they decided that they did not want to be like Europe, that part of their identity, and of meaningful life generally, was in close proximity to wildness. Thus was born the metaphorical frontier—the myth of the cowboy, the vast landscape of the possible, the ideal of the rugged individual—that defines American culture to this day. Adams's work grew to maturity alongside this metaphor and derives its power by eliciting the nostalgia for un-tamed wilderness at its core.

The idea of the frontier is not just quaint provincialism. It is often spoken of as such, especially in Europe, where the mythological na-ture of the American West has always been easier to discern than it is here and is often viewed with suspicion. I first saw this idea expressed in a lengthy article on America in the magazine *Stern* when I was a soldier stationed in Germany in the early 1970s. Such articles are ap-pearing with increasing frequency nowadays as the cold war recedes into history. But the perception is incorrect. While the confluence of cultural forces that generated Adams's images is uniquely American, the images themselves are not. The longing for a frontier seems to lie deep in the human soul, and people from different parts of the world and with different cultural backgrounds understand it quickly and intuitively. In no country does one have to dig very deep to find an appreciation of, and identification with, wildness. Adams's work travels well for this reason and has universal appeal.

The idea of science as a great frontier is similarly timeless.[3] While there are clearly many nonscientific sources of adventure left, science is the unique place where genuine wildness may still be found. The wildness in question is not the lurid technological opportunism to which modern societies seem so hopelessly addicted, but rather the pristine natural world that existed before humans arrived—the vast openness of the lone rider splashing across the stream with three pack animals under the gaze of mighty peaks. It is the choreography of ecologies, the stately evolution of minerals in the earth, the mo-tion of the heavens, and the birth and death of stars. Rumors of its death, to paraphrase Mark Twain, are greatly exaggerated.

My particular branch of science, theoretical physics, is concerned with the ultimate causes of things. Physicists have no monopoly on ultimate causes, of course, for everyone is concerned with them to some extent. I suspect it is an atavistic trait acquired long ago in

Africa for surviving in a physical world in which there actually are causes and effects—for example between proximity to lions and being eaten. We are built to look for causal relations between things and to be deeply satisfied when we discover a rule with cascading implications.[4] We are also built to be impatient with the opposite— forests of facts from which we cannot extract any meaning. All of us secretly wish for an ultimate theory, a master set of rules from which all truth would flow and that could forever free us from the frustration of dealing with facts. Its concern for ultimate causes gives theoretical physics a special appeal even to nonscientists, even though it is by most standards technical and abstruse.

It is also a mixture of good news and bad news. First you find that your wish for an ultimate theory at the level of human-scale phenomena has been fulfilled. We are the proud owners of a set of mathematical relationships that, as far as we know, account for everything in the natural world bigger than an atomic nucleus. They are very simple and beautiful and can be written in two or three lines. But then you find that this simplicity is highly misleading— rather like those inexpensive digital wristwatches with only one or two buttons. The equations are devilishly difficult to manipulate and impossible to solve in all but a small handful of instances. Demonstrating that they are correct requires arguments that are lengthy, subtle, and quantitative. It also requires familiarity with a huge body of work done after the Second World War. While the basic ideas were invented by Schrödinger, Bohr, and Heisenberg in the 1920s, it was not until powerful electronic computers were developed and armies of technically competent people were generated by governments that these ideas could be tested quantitatively against experiment over a wide range of conditions. Key technical developments, such as the purification of silicon and the perfection of atomic beam machines, were also important. Indeed, we might never have known for certain that the whole thing was correct had it not been for the cold war and the economic importance of electronics, radar, and accurate timekeeping, which made financing easy on various ostensibly practical grounds.

Thus eighty years after the discovery of the ultimate theory we find ourselves in difficulty. The repeated, detailed experimental confirmation of these relationships has now officially closed the frontier of reductionism at the level of everyday things. Like the closing of the American frontier, this is a significant cultural event, causing thoughtful people everywhere to debate what it means for the future of knowledge. There is even a best-selling book exploring the premise that science is at an end and that meaningful fundamental discovery is no longer possible. At the same time, the list of even very simple things found "too difficult" to describe with these equations continues to lengthen alarmingly.

Those of us out on the real frontier listening to the coyotes howl at night find ourselves chuckling over all this. There are few things a real frontiersman finds more entertaining than insights about wilderness from people back in civilization who can barely find the supermarket. I find this moment in history charmingly similar to Lewis and Clark's wintering on the Columbia estuary. Through grit and determination their party had pushed its way across a continent, only to discover that the value had not been in reaching the sea but in the journey itself. The official frontier at that time was a legal fiction having more to do with property rights and homesteading policy than a confrontation with nature. The same is true today. The real frontier, inherently wild, may be found right outside the door, if one only cares to look.

Despite being a wild place, the frontier is regulated by laws. In the mythical old West the law meant the force of civilization in a land where there was none, and it was often enforced by some heroic figure holding back the wildness of human nature through strength of will. A man had a choice of whether to obey this law or not, but he stood a good chance of getting gunned down if he did not. But there are natural laws as well, relationships among things that are always true regardless of whether people are present to observe them. The sun rises every morning. Heat flows from hot things to cold ones. Herds of deer spotting cougars always dash away. These are the exact opposite of laws of myth, in that they flow out of wildness and constitute its essence rather than being a means for its containment. Indeed, describing these things as laws is somewhat misleading, for it implies a kind of statute that otherwise willful natural things choose to obey. This is not correct. It is a codification of the way natural things are.

The important laws we know about are, without exception, serendipitous discoveries rather than deductions. This is fully compatible with one's everyday experience. The world is filled with sophisticated regularities and causal relationships that can be quantified, for this is how we are able to make sense of things and exploit nature to our own ends. But the discovery of these relationships is annoyingly unpredictable and certainly not anticipated by scientific experts. This commonsense view continues to hold when the matter is examined more carefully and quantitatively. It turns out that our mastery of the universe is largely a bluff—all hat and no cattle. The argument that all the important laws of nature are known is simply part of this bluff. The frontier is still with us and still wild.

The logical conflict between an open frontier on the one hand and a set of master rules on the other is resolved by the phenomenon of emergence. The term *emergence* has unfortunately grown to mean a number of different things, including supernatural phenomena not regulated by physical law. I do not mean this. I mean a physical principle of organization. Human societies obviously have rules of organization that transcend the individual. An automobile company, for example, does not cease to exist if one of its engineers gets run over by a truck. The government of Japan does not change very much after an election. But the inanimate world also has rules of organization, and they similarly account for many things that matter to us, including most of the higher-level physical laws we use in our daily lives. Such commonplace things as the cohesiveness of water or the rigidity of steel are simple cases in point, but there are countless others. Nature is full of highly reliable things that are primitive versions of impressionist paintings. A field of flowers rendered by Renoir or Monet strikes us as interesting because it is a perfect whole, while the daubs of paint from which it is constructed are randomly shaped and imperfect. The imperfection of the individual brush strokes tells us that the essence of the painting is its organization. Similarly, the ability of certain metals to expel magnetic fields exactly when they are refrigerated to ultralow temperatures strikes us as interesting because the individual atoms out of which the metal is made cannot do this.

Since principles of organization—or, more precisely, their consequences—can be laws, these can themselves organize into new laws, and these into still newer laws, and so on. The laws of electron motion beget the laws of thermodynamics and chemistry, which beget the laws of crystallization, which beget the laws of rigidity and plasticity, which beget the laws of engineering. The natural world is thus an interdependent hierarchy of descent not unlike Jonathan Swift's society of fleas:

> *So, naturalists observe, the flea*
> *Has smaller fleas that on him prey;*
> *And these have smaller still to bite 'em*
> *And so proceed* ad infinitum.

This organizational tendency is so powerful that it can be difficult to distinguish a fundamental law from one of its progeny. The only way we know that the behavior of cats is not fundamental, for example, is because cats fail to work when pushed beyond their proper operating limits, so to speak. Similarly, the only way we know atoms are not fundamental is that they come apart when caused to collide at great speed. This principle continues down to smaller and smaller scales: the nuclei from which atoms are made come apart when caused to collide at greater speed, the parts liberated from the nucleus come apart at even greater speeds, and so forth. Thus the tendency of nature to form a hierarchical society of physical laws is much more than an academic debating point. It is why the world is knowable. It renders the most fundamental laws, whatever they are, irrelevant and protects us from being tyrannized by them. It is the reason we can live without understanding the ultimate secrets of the universe.

Thus the end of knowledge and the closing of the frontier it symbolizes is not a looming crisis at all, but merely one of many embarrassing fits of hubris in civilization's long history. In the end it will pass away and be forgotten. Ours is not the first generation to struggle to understand the organizational laws of the frontier, deceive itself that it has succeeded, and go to its grave having failed. One would be wise to be humble, like the Irish fisherman observing quietly that the sea is so wide and his boat so small. The wildness we all need to live, grow, and define ourselves is alive and well, and its glorious laws are all around.

Living with Uncertainty

Fast is fine, but accuracy is everything.

Wyatt Earp

MY GENETICIST COLLEAGUE DAVID BOTSTEIN OFTEN BEGINS
lectures by explaining that the essence of biology is living with uncertainty. He especially emphasizes this to audiences of physicists,
because he knows they have a hard time with the concept and will
misinterpret much of what he says unless alerted to the issue ahead
of time. He has never revealed to me how he thinks about such audiences, but I happen to know that most biologists consider the
physicists' obsession with certainty and correctness to be exasperat-
ingly childish and evidence of their limited mental capacities. Physicists, in contrast, consider tolerance of uncertainty to be an excuse
for second-rate experimentation and a potential source of false
claims. These cultural differences have their roots in the historical
development of the two sciences (physics and chemistry evolved together with engineering, while biology came from agriculture and
medicine), and they mirror differences in our society generally about
what is and is not real and important. But because of them there is

relatively little useful communication between physicists and biologists at the moment.

A version of this communication problem comes up now and then in conversations with my wife, typically over money. She usually begins by casually suggesting some horrendously expensive purchase she cannot make on her own. I then ask her questions that I think get to the bottom of things, such as how much interest we will be paying or what the impact will be on our total cash flow. She responds that I am impossible because I always want to see things as black and white, never gray. I explain that I am just trying to solve the problem. She counters that I am oversimplifying. The world is nuanced, she says, not always clear-cut, and my insistence on stuffing things into categories and boxes is simply unreal. I respond that there is nothing unreal about avoiding jail and bankruptcy. The duration of this existential interchange depends on how much money is involved, but it eventually ends with some sort of compromise. Our argument is, of course, not about worldviews and reality at all but control of resources. I am the moralist in the family, so naturally I tend to lose more often than I win.

Physical scientists do not like absolute pronouncements about what is and is not true. We know that measurements are never perfect and thus want to know *how* true a given measurement is. This is a good practice, for it keeps everyone honest and prevents research reports from degenerating into fish stories. Our lofty attitude, however, belies something considerably easier to understand: the impulse to measure things accurately is the same as the impulse to make do-it-yourself repairs. The real allure is not high ideals at all but shiny, complex machines bristling with wires and dials, and staying up all night drinking coffee and manning the computer while the stereo blasts rock-and-roll in the background. It is monster X-ray tubes, smoking soldering irons, nuclear reactors with holes in them for neutrons to come out, highly dangerous chemicals, and helpful signs saying things like, "Do not look into the laser with your remaining good eye." It is also fundamentally a matter of problem-solving strategy, the notoriously gender-linked personality trait that is the source of all those jokes about wives who cannot read maps and husbands who refuse to ask for directions.[1] It is why buildings and academic majors at the Massachusetts Institute of Technology have numbers rather than names. Accurate measurement is simply natural behavior for people who see nothing strange in creating building ten, building thirteen, and course eight. I think all of this is mighty fine myself, but it is not for everybody.

One of the things we technological people find gratifying about giving in to this impulse is the world of meaning revealed by increasingly accurate measurement. For example, at an accuracy of one part in one hundred thousand, one discovers that the length of a brick is not the same from one day to the next. A check of environmental factors reveals this to be due to variations in temperature, which cause the brick to expand and contract slightly. The brick has become a thermometer. This observation is not silly, since thermal expansion is the principle behind all common thermometers.[2] A weight measurement to similar accuracy shows no such variations—one of many observations leading to the concept of inviolability of mass. But at an accuracy of one part in one hundred million, the weight of the brick becomes slightly different from one laboratory to the next. The brick is now a gravity meter, for this is an effect of slight variations in the force of gravity due to differing densities of rock immediately below the earth's surface.[3] Attaching the brick to a string and suspending it from the ceiling turns the brick into a pendulum, whose swing rate is also a measure of the force of gravity. The extreme stability of the swing is the principle behind pendulum regulation of mechanical clocks.[4] If the ceiling is high, the mass is large, and the swivel is outfitted with a little electric amplifier to prevent the pendulum from running down, the plane of the swing may be observed to rotate in response to the rotation of the earth, the rate of this rotation being a measure of the latitude.[5] Nontechnical people put up with this measurement obsession, which they otherwise find annoying, because of the useful new technologies it generates.

Physical scientists, on the other hand, tend to see the matter morally. They orient their lives around the assumption that the world is precise and orderly, and that its occasional failure to conform to this vision is a misperception brought about by their not having measured sufficiently accurately or thought sufficiently carefully about the results. This sometimes has bittersweet consequences. My brother-in-law the divorce attorney says that his most exasperating clients are Silicon Valley engineers, who typically want to just write down the family assets, divide them equally, shake hands, and be done with it. He has to patiently explain that it is not that easy— that people often lie and manipulate in stressful situations, that one can sometimes deceive oneself, that the value of the assets is not absolute, that there is horse-trading to be done, that there will be messy contractual obligations left over, and so forth. This does not mean that the simpler view is wrong, merely that it is not always practical.

Over the past three centuries, obsessive attention to detail has slowly revealed that some physical quantities are not only accurately reproducible from one experiment to the next but are completely universal. It is hard to overstate how astonishing and disturbing this is. The extreme reliability and exactness of these quantities elevates their status from mere useful fact to a kind of moral certainty. Many people feel uncomfortable thinking of numbers in moral terms, but they should not. If I hit a dog with my car going forty miles per hour it has different implications than if I hit the dog going one mile per hour. The more carefully these quantities were measured, the more accurately their universal values became known, even as the limits of technical capability were pushed back in breathtaking ways, a process that continues today. The deeper meaning of these discoveries is still being debated, but everyone agrees that they are important, for such certainty is uncommon in nature and demands explanation.

A familiar example of such a universal quantity is the speed of light. In the late nineteenth century there was increasing interest in measuring the motion of the earth in its orbit around the sun by its effect on the light propagation speed seen by an observer on earth. This was a daunting technical challenge at the time, since it required measuring the speed of light to an accuracy of one part in a billion. How this was accomplished is a wonderful story told over and over again around the campfires of physics, but let us say for the present purposes that it was done with mirrors.[6] By 1891 it had become clear that the effect was at least a factor of two smaller than it should have been based on an analogy with sound and the known speed of earth in its orbit. By 1897 this had improved to a factor of forty, a disparity too great to be dismissed as irrelevant or an experimental artifact. The expected modification of the speed of light due to the earth's motion did not exist. This finding eventually led Albert Einstein to conclude that the speed of light is fundamental and that moving bodies must gain mass as their speed increases.

The existence of universal quantities that can be measured with certainty is the anchor of physical science. This essential truth is sometimes easy to forget, for the fundamentals of physics have been with us so long that many of them have ossified into clichés. But despite how one may feel about their message, the postmodernist philosophers have correctly and insightfully understood that scientific theories always have a subjective component that is as much a creation of the times as a codification of objective reality.[7] Otto von Bismarck's famous quip, "Laws are like sausages—it is best not to see them being made," applies even more brilliantly to scientific theories, or so is my experience. As in all other human activities, it is necessary in science to take stock every now and then and reevaluate what one deeply understands and what one does not. In physics, this reevaluation nearly always comes down to precision measurement. Deep inside every physical scientist is the belief that measurement accuracy is the only fail-safe means of distinguishing what is true from what one imagines, and even of defining what true means. There is no need to have postmodernist anxieties about a universal number measured to one part in ten billion.

When physicists get together at parties to talk in uninhibited ways about things that matter to them, one of their favorite subjects is a famous lecture delivered by Irving Langmuir, the inventor of the modern tungsten-filament light bulb, on the subject of pseudoscience.[8] This lecture contains delicious case histories of scientific fakeries and swindles, but its greater importance lies in its central message: in physics, correct perceptions differ from mistaken ones in that they get clearer when the experimental accuracy is improved. This simple idea captures the essence of the physicist's mind and explains why they are always so obsessed with mathematics and numbers: through precision, one exposes falsehood.

A subtle but inevitable consequence of this attitude is that truth and measurement technology are inextricably linked. Exactly what you measure, how the machine works, how one decimates the errors, what uncontrolled factors set the reproducibility ceiling, and so forth matter more in the end than the underlying concept. In public we speak about the inevitability of these universal quantities, but in private we consider it unprofessional to talk about what ought to be universal in the same way we consider it unprofessional to talk about how much money one ought to make on stocks. You have to actually do the experiment. This practice may seem like the worst kind of pedantry, but it is really just common sense. Time and again things people thought were universal turned out not to be, and other things people thought varied actually didn't. Accordingly, when we speak of universal quantities we really mean the experiments that measure them.

The handful of experiments that are enormously accurate has, for this reason, a significance in physics greatly exceeding its size. There are between ten and twenty of these special experiments, depending on how one counts, and they are all revered.[9] Most of these special experiments are unfamiliar to nonexperts. There is the speed of light in vacuum, known now to an accuracy of better than one part in ten trillion. There is also the Rydberg constant, the number characterizing the quantization of light wavelengths emitted from dilute atomic vapors and responsible for the astonishing reliability of atomic clocks, known to an accuracy of one part in one hundred trillion. Another example is the Josephson constant, the number relating the voltage applied to a certain kind of metallic sandwich to the frequency of radio waves it emits, known to an accuracy of one part in one hundred million. Yet another is the von Klitzing resistance, the number relating the electric current forced through a specially designed semiconductor to the voltage induced at right angles by means of a magnet, known to an accuracy of one part in ten billion.

Paradoxically, the existence of these highly reproducible experiments leads us to think in two mutually incompatible ways about what is fundamental. One is that exactness reveals something about the primitive building blocks out of which our complicated, uncertain world is made. Thus we say that the speed of light is constant because it just is, and because light is not made of anything simpler. This thought process leads us to render these highly accurate experiments down to a handful of so-called "fundamental" constants. The other is that exactness is a collective effect that comes into existence because of a principle of organization. An example of the latter is the relationship between pressure, volume, and temperature of a gas such as air. The universal number characterizing the dilute gas law is known to an accuracy of one part in one million, yet it acquires huge errors in gas samples that are too small, and ceases to be measurable at all at the level of a few atoms. The reason for this size sensitivity is that temperature is a statistical property, like the market demand for houses, which requires a large sample to be defined. There is no way to reconcile these two ideas; they are exact opposites. Yet we use the word *fundamental* to describe both.

This dilemma is, of course, artificial. Only the collective idea is right. This is not obvious, and would even be denied vehemently by some physicists, but it becomes clear after one thinks critically about the experiments themselves and how they work.

Collective exactness tends to be a tough concept for nonscientists to grasp, but it shouldn't be. There are many familiar examples of it in daily life—for example, commuting. The sun comes up in the morning, and this is a reliable truth having to do with the primitive motion of the earth, the huge heat capacity of the sun, and so forth. But there is another, equally important, truth that the expressways and trains are always jammed with commuters at certain times of day, and moreover that the number of commuters is predictable from one hour to the next. It is certainly imaginable that all these commuters might get the stomach flu on the same day and stay home, but it is so unlikely as to be effectively impossible. The commute condition is a simple, reliable phenomenon that emerges out of complex decisions made by a large number of individuals as they go about their lives. It is not necessary to know what various individuals had for breakfast, where they work, what the numbers and names of their children are, and so forth, in order to appreciate that it's hell out there at 8:15 in the morning. Commuting traffic, like the behavior of the dilute gas, is a collective certainty. Whether it is as reliable as the sun rising must ultimately be determined by experiment, but my experiences commuting say it is.

A nice example of a collective effect masquerading as a reductionist one is the quantization of atomic spectra. Light is emitted from dilute atomic vapors with special wavelengths so insensitive to outside influences that they can be used to make clocks accurate to one part in one hundred trillion. But these wavelengths have a detectable shift at one part in ten million—ten million times larger than the timing errors of the clock—which should not have been present in an ideal world containing nothing but the atom.[10] Difficult but well-controlled calculations then revealed this shift to be an electrical effect of the vacuum of space not very different from what an electron encounters as it moves about inside a piece of metallic wire or a computer chip. The ostensibly empty vacuum of space, in other words, is not empty at all but full of "stuff." Its sympathetic motion when matter passes by changes the matter's properties slightly, just the way sympathetic motion of the electrons and atoms in a piece of window glass modifies the properties of light as it passes through, causing it to refract. The extreme reproducibility and reliability of these atomic experiments are thus crucially dependent on the uniformity of this "stuff," the cause of which is unknown. Identifying a plausible explanation for this uniformity is one of the central problems of modern physics and the chief objective of inflationary cosmologies—theories of the universe that are inherently emergent.[11] So even the constancy of atomic spectra actually has collective origins, the collective phenomenon in this case being the universe itself.

A much more immediate and troubling case of collectivism is the determination of the electron charge and Planck's constant by means of macroscopic measurements. The electron charge is the indivisible unit of electricity. Planck's constant is the universal relation between momentum and length that characterizes the wave nature of matter. Both are highly reductionist concepts, and both are traditionally determined using huge machines that measure properties of individual electrons ripped off of atoms. But their most accurate determination turns out to come not from these machines at all but simply from combining the Josephson and von Klitzing constants, the measurement of which requires nothing more sophisticated than a cryogenic refrigerator and a voltmeter.[12] That this was so was a great surprise when it was discovered, because the samples on which the Josephson and von Klitzing measurements are performed are highly imperfect. Chemical impurities, misplaced atoms, and complex atomic structures such as grain boundaries and surface morphologies are all plentiful and should have been able to disrupt the measurements at the reported level of accuracy. The fact that they do not *proves* that powerful principles of organization are at work.

One of the reasons physicists so rarely talk about the collective nature of measurements of fundamental constants is that it has such deeply troubling implications. Insofar as our knowledge of the physical world rests on experimental certainty, it is logical that we should associate the greatest truth with the most certain measurement. But this would seem to imply that a collective effect can be more true than the microscopic rules from which it descends. In the case of temperature, a quantity that never had a reductionist definition in the first place, this conclusion is easy to understand and accept. Every physical scientist understands that the tendency of heat to flow from hot things to cold ones is very general and would not be affected if one were to change the microscopic aspects radically—for example, by doubling the masses of all the atoms in the universe—so long as the system did not get small. But the electron charge is another matter. We are accustomed to thinking of this charge as a building block of nature requiring no collective context to make sense. The experiments in question, of course, refute this idea. They reveal that the electron charge makes sense *only* in a collective context, which may be provided either by the empty vacuum of space, which modifies this charge the same way it modifies atomic wavelengths, or by some matter that preempts the vacuum's effects. Moreover, the preemptive ability of matter requires the organizational principles at work there to be the same as those at work in the vacuum, since otherwise the effects would be miracles.

The electron charge conundrum, as it turns out, is not unique. *All* the fundamental constants require an environmental context to make sense. As a practical matter, the distinction between reductionist and emergentist quantities in physics does not exist. It is simply an artistic invention of humans, rather like the genders we sometimes assign to inanimate objects.

The idea of certainty emerging through organization is deeply embedded in the culture of modern biology, and is one of the reasons my colleagues in the life sciences are so eager to declare their tolerance of uncertainty. It shows they know the scoop. What they actually mean by such statements is that microscopic uncertainty does not matter, because organization will create certainty later on at a higher level. Another reason, of course, is that they want to loosen up the purse strings, the political strategy employed by my wife in those spending discussions. In neither case should the tolerance be taken at face value. Were it really the case that the essence of biology is uncertainty, then biology would not be science.

In physics, in contrast, the profound ideological disagreement on where certainty comes from, and what it means, remains unresolved. Instead, we agree not to talk about it. This compromise calls to mind Deng Xiaoping's famous remark that it does not matter whether a cat is black or white as long as it catches mice.[13] It is not uncommon for a committed reductionist to dismiss the evidence of the fundamental nature of collective principles on the grounds that there actually is a deductive path from the microscopic that explains the reproducibility of these experiments. This is incorrect. The microscopic explanation of temperature, for example, has a logical step called the postulate of equal a priori probability—a kind of Murphy's law of atoms—that cannot be deduced and is a succinct statement of the organizing principle responsible for thermodynamics.[14] The ostensibly deductive explanations of the Josephson and von Klitzing effects always have an "intuitively obvious" step in which the relevant organizational principles are assumed to be true. They actually are true, of course, so the reasoning is correct, but not necessarily in the sense the reasoner intended. In deference to reductionist culture, theorists often give these effects fancy names, which, on close inspection, are revealed to be nothing more than synonyms for the experiments themselves. In neither case was the great accuracy of the measurement predicted theoretically.

Like other things one does not talk about, unclear thinking about what is fundamental can come back to haunt us later on. Its most insidious effect is to lead us out into the desert by inducing us to search on smaller and smaller scales for meaning that is not there. I have a big problem with this—no doubt for cultural reasons. In the arid part of the world in which I grew up we take the desert seriously.

One of my great-grandfathers came to California by the Santa Fe Trail as a teenager and recorded his experiences along the way in a diary. According to this diary, he and his party had an extremely close shave somewhere in New Mexico. They had pulled into a small town to pick up supplies and water and ask for guidance on how to cross the desert. Upon receiving directions they struck out and, in two days, reached the first water hole and found it dry. Then they pushed on two more days to the second water hole and found it dry, too. Then they pushed on an additional two days and found another dry hole. At this point it became clear that the people back in that town had intended to kill them, so the party held a conclave and resolved on desperate measures. The men unhitched the horses from the wagons, left the women and children in the desert with all the supplies, rode back into town, shot it up, and brought back water. The story obviously had a happy ending, since I am here.

Despite the evidence that even physicists, ostensibly the most logical of scientists, can draw invalid conclusions from precise measurements, precision and certainty will continue to be scientific values that we cannot live without, because striving for certainty in measurement and interpretation is the only foolproof mechanism we have for revealing the principles of organization regulating the universe. Technical knowledge is just as susceptible to political whim as any other kind of knowledge, and it is only the anchor of certainty that gives science its special status and authority. The striving for certainty is not an anachronism of a bygone era promoted by Luddite physicists but the moral core of science. It is like old-time religion—occasionally annoying and tiresome but never irrelevant. All of us, and perhaps even all living beings, use the especially reliable things that nature sees fit to reveal to us as beacons to navigate through an otherwise uncertain world. As with any other aspect of life, one of the worst things a body can do is to allow this system to weaken by miscategorizing a falsehood as a truth. The consequence will be that the system fails at the crucial moment one needs it most, causing one to lose one's way.

THREE

Mount Newton

Nature's laws are the invisible government of the earth.

Alfred A. Montapert

IN 1687, ISAAC NEWTON CHANGED HISTORY BY LAYING down in the *Principia* the scientific case for universal physical law.[1] Regularity in the natural world had been well understood since ancient times, and Renaissance figures like Galileo, Kepler, and Tycho Brahe had recently refined and quantified this knowledge through careful experimental observation. But Newton went beyond observation of regularity to identify mathematical relationships that were simple, applied always, and accounted for apparently unrelated behaviors simultaneously. Newton's laws of motion turned out to be so trustworthy that incompatibility with them soon became a reliable indicator of false observations. They found important applications in engineering, chemistry, and commerce and eventually became the logical basis for our entire technological world. Little wonder that Alexander Pope's famous eulogy still brings a tear to the eye:

Nature and Nature's laws lay hid in night.
God said, Let Newton be! and all was light.

Much creative energy has gone into testing and exploiting Newton's laws.

The great influence of Newton's treatise came not from its explanation of planetary orbits and the tides, which was very beautiful, but from its use of these things to demonstrate the legitimacy of the clockwork universe—the idea that things tomorrow, the day after, and the day after that are completely determined from things now through a set of simple rules and *nothing else.*[2] The stunning quantitative agreement between Newton's calculations and experimental observations of the planets left no doubt that his rules were correct for astronomical bodies, and that the mystery of the heavens had been solved. The simplicity of these rules, their reasonableness, and their compatibility with Galileo's terrestrial observations also suggested that they applied much more generally—that they were the machinery of the clock. This has been borne out by subsequent observations. In four centuries of careful experimentation the only documented failures of Newton's laws of motion have been at atomic-length scales, where the laws of quantum mechanics supplant them.

We know Newton's laws to be highly accurate because so much creative energy has gone into testing and exploiting them. There are

several classes of tests. One is the careful observation of the motion
of astronomical bodies. Newton's laws not only account for shapes
and histories of planetary orbits in detail but also correctly predict
the sun's effects on the orbit of the moon, the complex trajectories of
asteroids and comets,[3] and the stability of the asteroid belt. The apparent failure of Uranus to obey Newton's laws led to the discovery of
Neptune and then Pluto.[4] Another class of test consists in the study
and manufacture of accurate mechanical clocks, ranging from the
original Huygens pendulum clock and its progeny to the balance
wheel chronometer[5] and to the quartz oscillator used in modern
wristwatches.[6] Yet another class is based on the principle of the gyroscope and the technology of the gyrocompass and gyrostabilizer built
upon it.[7] Newtonian ideas are used in designing machinery and the
earthquake stability of tall buildings, and are implicit in laws of electricity that lead to power transmission, computers, and radio.

Despite the successes of Newton's laws and the engineering advances they made possible, many people still find the clockwork
universe difficult to accept. It flies in the face of our commonsense
understanding of the complexity of nature and our belief that the
future is not completely predestined but depends on how we choose
to behave. It also seems to be inconsistent with everyday experience
and to have moral implications that are not right. It can, for example, become an excuse to do anything you wish to other people and
create dangerous things as you see fit because nature is, after all, just
mechanical. It also can legitimize a bogus faith in logic. I first heard
this latter idea articulated by my father long ago during a dinner-
table discussion about predestination. At one point he became exas-
perated with the barrage of ignorant statements about reality from
the kids and explained, barely controlling himself, that logic was the
systematic method of committing error. Now that I am older I understand what he meant. He knew through painful experience in his
law practice that human beings reason by analogy. When we say

something is unreasonable, we usually mean it is not suitably analogous to things we already know. Pure logic is a superstructure built on top of this more primitive reasoning facility and is thus inherently fallible. Unfortunately, we need to be most logical precisely when it is most difficult—when confronted with something new that is not analogous to anything we already know. The ability to do this intensely for long periods of time is what distinguishes the Isaac Newtons and Albert Einsteins from the rest of us. So on this matter my father was right, but only partly. Logic sometimes can, and must, be believed. The material evidence for the clockwork universe has grown over the centuries to become overwhelming. One must look somewhere other than a failure of this idea for the answers to the mysteries of life.

The moral conundrum of material determinism was even more troublesome in the seventeenth century, when physics was being invented, than it is today. In 1633 Galileo Galilei was brought to trial before the Italian Inquisition for violating a 1616 edict against promoting the cosmology of Copernicus. He was found "vehemently suspected of heresy," a judgment slightly less severe than actual heresy, and was forced to publicly recant his belief that the earth moves about the sun.[8] Like many great scientists, Galileo was a rebellious individual. He had left university without a degree in order to pursue his own intellectual agenda of measuring things rather than just thinking about them. His career was dazzlingly successful. We know Galileo today mostly for his invention of the astronomical telescope and the discoveries he made with it, such as sunspots and the moons of Jupiter,[9] but his deeper contribution was articulating the fundamental limitations of Aristotle's discursive approach to science and advocating the need for mathematical precision. The Book of Nature, he wrote in *The Assayer* in 1623, ". . . is written in the language of mathematics."[10] Unfortunately, Galileo's deterministic worldview, forcefully argued in that book, left no room for divine intervention and, perhaps even worse, implicitly promoted the idea that divine things could be understood and mastered by humans. In 1625 he was secretly denounced to the Inquisition for the threat to Eucharistic theology, in particular the doctrine of transubstantiation, in *The Assayer*, which, ironically, he had dedicated to his good friend Cardinal Maffeo Barberini, on the occasion of his election as Pope Urban VIII in 1623.[11] The matter came to a head in 1632 when Galileo published his great work, *Dialogue Concerning the Two Chief World Systems*, a brilliant and devastating scientific attack on the Ptolemaic universe.[12] On advice that its arguments were so lucid and persuasive that it was more dangerous than Calvin and Luther combined, the Pope ordered that publication of the book cease and that Galileo be brought to trial. He was found guilty and sentenced to house arrest in Arcetri, a small village outside Florence, where he remained for eight years until his death.

Without Galileo, Newton's work would have been unthinkable. Nearly all of Newton's essential physical ideas—and the experiments that backed them up—were originally due to Galileo. It was Galileo who first realized that objects did not require an external agent to move them, as Aristotle had thought, but instead moved at constant speed on straight-line trajectories unless acted upon from without. Galileo also invented the idea of velocity as a vector, a quantity with both magnitude and direction. He invented the idea of inertia, the natural resistance of a body to changes in its motion, and was the first to identify the agent for modifying motion as force, a thing that changes the velocity additively from one moment to the next, so that the velocity two seconds from now is the velocity now plus a small increment that depends on the magnitude of the force.

Isaac Newton nonetheless receives the lion's share of the credit for inventing modern physics because he discovered a way to synthesize all these ideas into a seamless mathematical whole. He was born on Christmas day 1642, the year Galileo died.[13] Like Galileo, Newton was a rebellious individual disinclined to trust authority. In the margin of one of his Cambridge notebooks is scribbled in Latin, "Amicus Plato, amicus Aristoteles; magis amica Veritas." (Plato is my friend, Aristotle is my friend, but truth is a better friend.) Like many motivated young people of his day, he was fascinated by the new astronomy and had read Galileo and Kepler extensively. We owe Newton's discoveries in no small measure to the Great Plague, from which he hid at his home in Lincolnshire between 1665 and 1667. While there, presumably with time on his hands, he invented the infinitesimal calculus, the key breakthrough required for explaining Kepler's observations about planetary orbits—their planarity, their perfect elliptical shape with the sun at one focus, their miraculous accelerations and decelerations that caused equal areas of the ellipse to be swept out in equal times, and the exact mathematical relationship between the size of the orbit and its period. With the notation of calculus, Newton was able to write down Galileo's rules of motion as simple, precise equations, which could then be solved to obtain an exact description of a body's motion in response to the forces acting upon it. With this mathematical technology and one further assumption— that the force of gravity weakened in a certain way with distance—he was able to prove that Kepler's observations actually followed from Galileo's rules and were not independent phenomena.[14] This, in turn, enabled him to argue from the extreme accuracy of Kepler's observations that Galileo's rules were exact. Galileo had missed this point entirely. He had ignored Kepler's laws, which were discovered in his lifetime, and had considered the whole idea of universal gravitation "occult." Fate had apparently ordained that Galileo should lead his people to the Promised Land but not enter in himself.

One of the greatest disservices we do to our students is to teach them that universal physical law is something that obviously ought to be true and thus may be legitimately learned by rote. This is terrible on many levels, the worst probably being the missed lesson that meaningful things have to be fought for and often require great suffering to achieve. The attitude of complacency is also opposite to the one that brought these beautiful new ideas into the world in the first place—indeed, what brings things of great importance into the world generally. The existence of physical law is, in fact, astonishing and should be just as troubling to a thinking person today as it was in the seventeenth century when the scientific case for it was first made. We believe in universal physical law not because it ought to be true but because highly accurate experiments have given us no choice.

For some reason I was recently seized with concern about this problem while on a car trip with my family. I asked my son, who was taking physics in high school, what the evidence was that Newton's laws were true. He is a sympathetic person, so he dutifully rose to the bait, bluffed valiantly, realized that what he was saying did not make any sense, twisted in the wind a bit, mumbled something I could not make out, and then fell silent. I clarified the question by asking him what the key experiments were. More silence. This happy moment was an effect of the universal gene parents have for giving their children reasons to hate them. I was fully aware that he did not know the answer, and I was trying to provoke a thoughtful discussion about planetary orbits—which I successfully did in the end. I am reasonably sure the outcome was positive, but one will only know for sure when negotiations begin for dividing up the estate.

Universal physical law is the iceberg of which the exact physical constant is the small part above water. Both are aspects of the same physical phenomenon, but physical law is the vastly more inclusive concept. In the Far East, where I travel frequently, I like to explain this using an analogy with the Theravada and Mahayana branches of Buddhism.[15] In the Theravada, one restricts one's attention to the conservative teachings of specific historical scholars. In the Mahayana, or "great vehicle," one considers not only these teachings but all of their many implications. A universal constant is a measurement that comes out the same every time. A physical law is a *relationship* between measurements that comes out the same every time. In the case of laws of motion such as Newton's, it is a relationship between measurements at different moments. Thus when we measure certain things now we need not measure them again in the future (assuming they remain undisturbed) because their values are predestined with certainty. In discussing laws we speak of exact equations instead of exact values, but the core idea is the same. Exactness is what counts. Like exact universal measurements, we tend to classify laws in our minds as either microscopic or collective in origin and use the word *fundamental* to describe both. As with constants, we find that the difference between these two classifications tends to melt away when the experimental facts are examined closely.

Over the years, as the list of successes of Newton's laws lengthened, there arose a speculative use of them very different from the original highly conservative one. The new strategy was to assume that Newton's laws were true in circumstances where one could not verify this directly, compute various physical properties based on this assumption, and then argue from agreement with experiment that the initial assumptions were correct. Thus, for example, the kinetic theory of gases assumes the gas to consist of atoms obeying Newton's laws with short-ranged repulsive forces that cause them to carom off each other like billiard balls. One then finds that the mythical atoms have a strong tendency to be scrambled into randomness by their collisions— as anyone who has played billiards knows well. This tendency is called the principle of chaos and is the origin of the unpredictability of the weather.[16] After scrambling, the chaotic swarm of billiard balls beautifully emulates the behavior of dilute gases, as well as corrections to ideal gas law as the gas density is increased, which come from the interatomic forces. Thus we say that the kinetic theory "explains" the ideal gas law—meaning that it accounts for the origin of the law.

But this reasoning has the obvious logical flaw that the behavior against which one tests the assumptions might be a universal collective phenomenon. In this case the measurement is fundamentally insensitive to microscopic assumptions, such as the existence of atoms, and therefore does not test them at all. It is a false syllogism: God is love, love is blind, Ray Charles is blind, therefore Ray Charles is God.[17] Unfortunately, this is precisely what happened in these theories. Newton's laws, as it turns out, are wrong at the scale of atoms.

Early in the twentieth century it was discovered that atoms, molecules, and subatomic particles are described by the laws of quantum mechanics—rules so different from Newton's that scientists struggled to find proper words to describe them. Newton's laws make profoundly false predictions at this scale, such as atoms having zero size and solids having huge heat capacities at zero temperature that they do not, in fact, have. A beam of helium atoms projected onto an atomically perfect solid surface does not bounce off in all directions, as Newton's laws predict, but diffracts into rainbows as a beam of light would do.[18] Atoms are not billiard balls at all but waves, as are their constituents, which bind together to form atoms the way waves of water bind to make a surge.[19]

Thus Newton's legendary laws have turned out to be emergent. They are not fundamental at all but a consequence of the aggregation of quantum matter into macroscopic fluids and solids—a collective organizational phenomenon. They were the first laws to be discovered, they brought the technological age into existence, and they are as exact and true as anything we know in physics—yet they vanish into nothingness when examined too closely. Astonishing as it may seem, many physicists remain in denial. To this day, they organize conferences on the subject and routinely speak about Newton's laws being an "approximation" for quantum mechanics, valid when the system size is large—even though no legitimate approximation scheme has ever been found. The requirement that Newton's laws emerge in the macroscopic limit was christened the *principle of correspondence* in the early days of quantum mechanics and was used as a constraint in working out the meaning of quantum measurement. The notoriously illogical (and partly wrong) ideas about quantum indeterminism still with us today are untidy consequences of this process. But the correspondence principle remains mathematically unprovable.

I first learned about the emergent nature of Newton's laws from P. W. Anderson's famous essay *More Is Different*. After thinking hard about why metals refrigerated to very low temperatures exhibit the bizarre exactnesses of superconductivity, Anderson realized that the central dilemma was precisely that of the correspondence principle. In other words, superconducting behavior reveals to us, through its exactness, that *everyday reality* is a collective organizational phenomenon.

So it seems that my poor son was intellectually mugged. I apologize, Todd.

FOUR

Water, Ice, and Vapor

Law is order, and good law is good order.

Aristotle

EVERY WEEKEND IN JANUARY, ARMIES OF TRUCKS DRIVE out onto the lakes of Minnesota in search of fish.[1] The drivers all understand the danger of this but are willing to risk it, for they are being driven bananas by winter and cannot resist the thought of all those waiting crappie, walleye, and jumbo perch. They invent all sorts of justifications for getting away, and even go so far as to claim that their wives love cleaning and preparing fish. It is a lie. Their wives hate fish and are always apprehensive, sometimes terrified, about these trips. They just put up with them because they have no choice. Considering the number of drivers, it is probably surprising that there are not more accidents. According to Tim Smalley, boat and water safety specialist at the Minnesota Department of Natural Resources, there were only 117 ice fatalities between 1976 and 2001, 68% of which involved a vehicle.[2] Evidently, ice is reliably strong and buoyant—at least in Minnesota's winters, and provided that the person testing its strength and buoyancy has not been drinking.

Not only Minnesotans with cabin fever but all of us entrust our lives to the solid state every day—from standing on ice to ordering peanuts at 40,000 feet—without thinking twice about it. We know empirically that matter sufficiently cold freezes, and that when it does, it universally and exactly acquires shape, form, and springy resistance to deformation. There is no possibility that the solid will suddenly lose its rigidity and betray us, even though a modest temperature rise—sometimes a fraction of a degree—can accomplish just this by causing it to melt. In the fiery hell of a furnace the metal may splash and play, but in our world it is sober and responsible.

The phases of matter—among them the familiar liquid, vapor, and solid—are organizational phenomena. Many people are surprised to learn this, since phases seem so basic and familiar, but it is quite true. Trusting the ice is less like buying gold than buying stock in an insurance company. If the organizational structure of the company were to fail for some reason, one's investment would vanish, for there is no physical asset underneath. Similarly, if the organization of a crystalline solid—the orderly arrangement of the atoms into a lattice—were to fail, the rigidity would vanish, since there is no physical asset underneath it either. The property we value in either case is the order. Most of us would prefer not to think we are entrusting our lives to an organization, but we do it every day. Without economies, for example, which are purely organizational phenomena, civilization would collapse and all of us would starve.

Ironically, the immense reliability of phase-related phenomena makes them the reductionists' worst nightmare—a kind of Godzilla set loose by the chemists to crush, incinerate, and generally terrorize their happy world. A simple, universal phenomenon one encounters frequently cannot depend sensitively on microscopic details. An *exact* one, such as rigidity, cannot depend on details at all. Moreover, while some aspects of phases are universal and thus easy to anticipate, others, such as which phase one gets under which circumstances, are not—

water being an especially embarrassing case in point. Ordinary water ice displays, at last count (the number keeps rising due to new discoveries) eleven distinct crystalline phases, not one of which was correctly predicted from first principles.[3] These phases, known as ice-I, ice-II, and so forth, are not to be confused with ice–9, the fictional weapon of mass destruction satirized in Kurt Vonnegut's novel *Cat's Cradle*.

Phases are a primitive and well-studied case of emergence, one that conclusively demonstrates that nature has walls of scales: microscopic rules can be perfectly true and yet quite irrelevant to macroscopic phenomena, either because what we measure is insensitive to them or because what we measure is overly sensitive to them. Bizarrely, both of these can be true simultaneously. Thus it is presently too difficult to calculate from scratch which crystalline phase of ice will form at a given temperature and pressure, yet there is no need to calculate the macroscopic properties of a given phase, since these are completely generic.

A measure of the seriousness of this problem is provided by the difficulty of explaining clearly how one knows phases to be organizational. The evidence always manages to be complicated, indirect, and annoyingly intermingled with theories—not unlike the evidence of product superiority in a commercial for soap or cars. The deeper reason in each case is that the logical link from the fundamentals to the conclusion is not very substantial. One thing we know for certain is that crystalline solids are ordered lattices of atoms—a fact revealed by their tendency to deflect X-rays through specific angles—while liquids and gases are not. We also know that systems with small numbers of atoms are motivated by simple, deterministic laws of motion and nothing else.[4] We also know that attempts to discover the scale at which these laws cease to work or are supplanted by others have failed. And finally, we know that elementary laws have the ability *in principle* to generate phases and phase transitions as organizational phenomena.[5] Thus when one strips away the unhelpful complexities, one is left with the following simple argument: microscopic laws are true and could plausibly cause phases; therefore we are sure they do cause them, even though we cannot prove this deductively. The argument is believable and, I think, correct, but it does have the strange effect of giving the word "cause" a meaning it does not customarily have. One could say that the laws of chemistry "caused" the destruction of Tokyo, but what really did it was Godzilla.

The believability of this argument gives phase organization an enormous importance that it would not otherwise have, for it is impossible to disguise the fact that phases are boring. From a practical standpoint there is not much difference between a law that emerges and a miracle that just is, but from a philosophical standpoint the difference is profound. One represents a world ruled by orderly hierarchical development, the other a world ruled by magic. The precedent of phases *proves* that at least some of the marvels of the world are organizational—and, in so doing, suggests that all of them are. It is one of the main reasons we tend to doubt supernatural causes for things until organizational causes have been ruled out experimentally.

There are lots of other everyday examples of exactness generated by phases. Liquids, for example, will not tolerate pressure differences of any kind between one point and another except those due to gravity. This is a general property of the liquid phase that does not depend on what the liquid is made of. It is not obvious, which is why the renowned Greek mathematician Archimedes screamed "Eureka!" upon discovering it and ran naked through the streets of Syracuse.[6] It is the principle behind the mercury barometer, the buoyancy of steel ships, and all hydraulic machinery. The liquid phase has an electronic version, the metallic phase, which will not tolerate voltage differences. This exact property of metals is the principle behind conduction of electricity in wires, as well as such practical rules such as not touching commercial radio towers while they are transmitting.

Both the liquid and metallic phases have special low-temperature versions, the superfluid and the superconductor, which have even more impressive exact behaviors.

The simplest prototype of emergent exactness, however, is the regularity of crystal lattices, the effect ultimately responsible for solid rigidity. The atomic order of crystals can be perfect on breathtakingly long scales—in very good samples, as many as one hundred million atomic spacings.[7] Atomic order was suspected as early as the seventeenth century as the cause of the simplicity and regularity of crystalline shapes,[8] but the degree of perfection was not known until X-ray crystallography was invented.[9] One infers the perfection of the ordering mostly from the exactness of X-ray reflections, although it is also detected indirectly through transport experiments such as conduction of electricity at low temperatures.

To appreciate the miracle of crystallization it is helpful to imagine a school with ten billion children. The recess bell rings, and the teachers line up the kids in rows upon rows on the gigantic playground in preparation for ushering them back into class. The kids have other ideas, however, for they have been wound up by their play and detest work. They fidget, pester each other, and run around in circles playing tag while the authorities struggle to achieve control. Without actually doing this experiment it is very hard to tell whether any long-range ordering pattern would materialize on the playground, for at the range of a few hundred children the pattern is highly flawed and arguably even nonexistent. But at the scale of one hundred thousand children the pandemonium of a single class might become irrelevant, allowing us to say that a one-hundred-kilometer crystal of children has formed.

It is not at all obvious that atoms in a crystal should order so well. For one thing, it does not always happen. Elemental helium, for example, remains liquid no matter how much its temperature is lowered, although it will crystallize when subjected to pressure.[10] Amorphous substances such as glass and plastic can be made to crystallize only with great difficulty and are usually found in a state of semipermanent frozen chaos.[11] It is still extremely difficult to predict which proteins will crystallize and which will not, despite its being a matter of immense importance to the modern drug industry.[12] Which things crystallize easily can be anticipated to some extent from their microscopic structure, but in the final analysis the perfection of crystal lattices just is. The last time the stock market crashed, the *Economist* explained that "shift happens." This is also how we explain the failure of crystals to form.

The most astonishing thing of all about crystalline ordering is that it remains exact when the temperature is raised. Temperature may be thought of as the amount of sugar our ten billion children have in their bloodstreams. Even in good crystals a given atom is always moving and thus always slightly off of its ideal lattice site at any given moment, this being the physical meaning of heat. The proof that this motion is present is that a fraction of the X-rays beamed into a sample are reflected with a small wavelength change, exactly as occurs when radar beams bounce off a moving airplane.[13] But astonishingly, this effect does not fuzz out the specific angles through which the X-rays are deflected; it only steals some of the deflected beam's intensity and redistributes it as a uniform background reminiscent of fogging in a photograph. This occurs because the location of one atom continues to predict the location of another—with some uncertainty—arbitrarily far away in the structure. The positional errors do not accumulate. This enables the line of children to look chaotic at the hundred-kid level but perfectly ordered at the million-kid level. In the liquid phase, in contrast, the deflected image does fuzz out because the positional errors do accumulate and do cause predictive power to be lost at sufficiently large distances. The lattice positions of a solid evidently have exact meaning even when the atoms are not exactly in them.

The exactness of the lattice registry on long-distance scales explains why melting is abrupt.[14] The ability of an atom to predict the position of another arbitrarily far away cannot be partially present any more than a person can be partially pregnant. When this predictability is present, simple logic tells us that the other properties one normally associates with solids, such as shape and elasticity, must also be present. These properties can therefore be lost only catastrophically. There are, unfortunately, constant misunderstandings about how much this exactness matters to the nature of the solid state. Most substances are not perfectly regular—even real metals, which owe much of their important engineering properties to structural and chemical imperfections.[15] An acrylic bowling ball dropped on one's foot may not be solid by a theorist's definition, but it certainly seems that way when you are sitting in Urgent Care waiting for the surgeon. But the abrupt transformation from solid to liquid that enables us to speak of these things as distinct phases requires ordering. In glasses or polymeric materials, such as a bowling ball, no abrupt phase transition occurs upon cooling; thus there is no meaningful experimental way to determine whether the substance is a solid or a highly viscous liquid.[16] The distinction is semantic—hence the intractability and bitterness of the disputes about it. In principle, a similar problem also occurs in impure crystals, but in practice, the disruption of the phase transition is usually too subtle to matter.

Anyone doubting the earnestness of phase transitions should be forced to winter in New England, a place notorious for capricious weather. When I was a graduate student I shared a house in the suburbs of Boston at the end of a cul-de-sac that was always difficult to deal with in snow emergencies. One day a winter storm blew in. It dumped snow from early morning until nine o'clock at night, whereupon the temperature suddenly warmed way up and it poured. The rain came down in tropical amounts and mixed with the snow already on the ground to make slush. This clogged the storm drains and filled the streets up to the curb. Then at about three in the morning, while everyone was asleep, an arctic front blew in from Canada, plunged the temperature back down below zero, and froze the street into a solid block of ice a foot thick. By morning, plowing was pointless, and cars unfortunate enough to have been parked on the street overnight were entombed. The city waited a week for a thaw that never came, then threw in the towel and dumped sand on top. This remained as a kind of dirty, slippery ice concrete until spring, when it finally melted away.

Once one knows what to look for, the organizational nature of phases other than the solid becomes easy to demonstrate. A collective state of matter is unambiguously identified by one or more behaviors that are exact in a large aggregation of the matter but inexact, or nonexistent, in a small one. Since the behavior is exact, it cannot change continuously as one varies external conditions such as pressure or temperature but can change only abruptly at a phase transition. One unambiguous signature of an organizational phenomenon is therefore a sharp phase transition. The transition itself, however, is only a symptom. The important thing is not the transition but the emergent exactness that necessitates it.

The melting and sublimation transitions of ice signal the demise of crystalline order and its replacement by a set of exact behaviors known collectively as hydrodynamics.[17] The laws of hydrodynamics amount to a precise mathematical codification of the things we intuitively associate with the fluid state, such as the meaningfulness of hydrostatic pressure, the tendency to flow smoothly in response to differences in pressure, and the rules of viscous drag. No one has ever succeeded in deducing these laws from first principles, although it is possible to make highly plausible arguments in many cases. The reason we believe them, as with most emergent things, is because we observe them. Like the laws of rigidity in solids, the laws of hydrodynamics become more and more exact as the length and time scales on which they are measured increase, and they fail in the opposite limit. The emergence of hydrodynamic law at long wavelengths is why compressional sound propagates universally in fluids and why the shear strength of a fluid is always exactly zero. The insensitivity of hydrodynamic principles to details allows deep-sea divers to continue speaking to each other, albeit with Donald Duck voices, when nitrogen in their breathing mixtures is replaced with helium.

Isotropic fluids are not just the opposite of solids but rather one of many possible alternatives to them. The most industrially significant of these are the liquid-crystal phases that constitute the active element of flat-panel computer monitors and cheap wristwatches.[18] These are characterized by an intolerance to shear stress, as in conventional liquids, but residual anisotropy that enables them to twist light polarization in response to small electrical signals. Another example is the hexatic phase, a state with fluidlike shear properties but sixfold orientational memory that forms when ordinary rare gas atoms condense on graphite.[19] (The hexatic phase is difficult to detect experimentally, so its existence is more controversial.) Another example is the "incompressible" phase, in which a fluid cannot transmit conventional sound, which occurs in magnetic fields. Yet another is the supersolid, a theoretical phase with shape rigidity that nonetheless flows, the experimental observation of which was recently reported.[20] These exotic phases are rare, but their existence is nonetheless important because it demonstrates the familiar solid, liquid, and gas to be special cases of something more general.

The exact property distinguishing the liquid phase of water from vapor is something considerably more subtle: the interface between them. Water and steam seem so different that it is hard to imagine that they would be difficult to tell apart, but they sometimes are. As one raises the steam pressure above a pot of boiling water (a side effect of which is to elevate the boiling temperature), the roiling surface becomes harder and harder to see and, at a critical pressure, disappears. Above this pressure the liquid and vapor have lost their separate identities and have merged into a single phase, the fluid, so there is no surface. The pressure at which the liquid and vapor merge is useful to engineers because the special expansion properties of steam they exploit to make engines are maximized there, but is otherwise unimportant. The emergent phenomenon distinguishing the liquid and vapor phases is thus not the development of order but the development of a surface. Like the lattice of a crystalline solid or the laws of hydrodynamics in the fluid, this surface and the rules for its motion become increasingly well defined at large distance and time scales but lose their meaning in the opposite limit. This is the effect that brings us clouds, rain, and the magnificent violence of the sea.[21]

By far the most important effect of phase organization is to cause objects to exist. This point is subtle and easily overlooked, since we are accustomed to thinking about solidification in terms of packing of Newtonian spheres. Atoms are not Newtonian spheres, however, but ethereal quantum-mechanical entities lacking that most central of all properties of an object—an identifiable position. This is why attempts to describe free atoms in Newtonian terms always result in nonsense statements such as their being neither here nor there but simultaneously everywhere. It is aggregation into large objects that makes a Newtonian description of the atoms meaningful, not the reverse. One might compare this phenomenon with a yet-to-be-filmed Stephen Spielberg movie in which a huge number of little ghosts lock arms and, in doing so, become corporeal. For this to occur, their number must be stupendously large. Merely bonding atoms together into a very large molecule will not suffice. Fullerenes, for example—soccer-ball-shaped molecules consisting of 60 or more carbon atoms—diffract very nicely and thus are still measurably quantum-mechanical.[22] But as the sample size grows to infinity, the distinction between the internal motions and the collective motion of the whole body becomes *qualitative*—and the latter acquires Newtonian reality.

The reason we get away with thinking of atoms as Newtonian is that an emergent phenomenon renders the mistake irrelevant. But it only does so for the motion of the object as a whole. The internal vibrational motions remain quite quantum.

Collective emergence of objects is the principle behind the phenomenon of superness that occurs in ultracold environments.[23] Like the comic book character Superman, superfluid helium can leap tall buildings in a single bound—or, more precisely, crawl up the walls of a beaker all on its own and escape. Unlike Superman, it has properties so strange and implausible that they could never have been accepted for publication at a pulp science fiction magazine. The viscosity of the superfluid is not just small but exactly zero, enabling it to pass through porous plugs as though they were not there and remain exactly stationary when its container is rotated. Superconductors similarly pass electric current with no resistive loss, and generate magnetism when rotated because the atomic nuclei move while the electrons do not.

Superfluidity and superconductivity are the fluid versions of ideal crystalline rigidity. This is not at all obvious, particularly since they appear to be special "quantum" phenomena that have no analogue in the Newtonian world, just as zero-temperature hydrodynamics does not, but this is incorrect. The tip-off is the exactness. By good fortune, the superfluid order, while exotic, is also simple and thus easy to understand. You might describe it as a tank of little ghosts drifting through each other but belonging by choice to the same political party—which one being immaterial, as long as it is the same. If one then perturbs the tank by forcing the political opinion to be one thing on the left side and another thing on the right side, the body politic of ghosts inside becomes stressed and responds with the mass migration we call superfluid flow.

Superfluid rigidity has enough in common with ordinary crystalline rigidity that one can draw useful analogies between the two.

Thus if one cools a rotating container of helium through its super-fluid transition, the fluid continues to rotate, but only in a lattice of tiny quantized vortex lines.[24] These are the fluid version of line defects in the crystal one could make by removing a thin pie slice with a knife and then squashing it together to reseal the place that was cut.[25] In the fluid there is no lattice to be defective, so the memory of the cut is preserved in a special persistent fluid flow about the line.

The crystalline and superfluid phases, and their attendant exact behaviors, are specific examples of an important abstract idea in physics called spontaneous symmetry breaking. It has uses ranging from engineering to the modern theory of the vacuum of space[26] and is even suspected of being relevant to life.[27] The idea of symmetry breaking is simple: matter collectively and spontaneously acquires a property or preference not present in the underlying rules themselves. For example, when atoms order into a crystal, they acquire preferred positions, even though there was nothing preferred about these positions before the crystal formed. When a piece of iron becomes magnetic, the magnetism spontaneously selects a direction in which to point. These effects are important because they prove that organizational principles can give primitive matter a mind of its own and empower it to make decisions. We say that the matter makes the decision "at random"—meaning on the basis of some otherwise insignificant initial condition or external influence—but that does not quite capture the essence of the matter. Once the decision is made, it becomes "real" and there is nothing random about it anymore. Symmetry breaking provides a simple, convincing example of how nature can become richly complex all on its own despite having underlying rules that are simple.

The existence of phases and phase transitions provides a sobering reality check on the practice of thinking of nature solely in terms of the Newtonian clockwork. Floating on the lakes of Minnesota and stretching into the sky in large cities are simple, concrete examples of how organization can cause laws rather than the reverse. The issue is not that the underlying rules are wrong so much as that they are ir-relevant—rendered impotent by principles of organization. As with human institutions, emergent laws are not trustworthy, and some-times hard to discern, when the organization is small, but they be-come more reliable as it grows in size and eventually become exactly true. This is why you can buy treasury bills with confidence or drive a truck out onto the ice with small risk. The analogy with human in-stitutions might seem a bit shaky in light of recent revelations of ac-counting swindles and financial collapse in large corporations, but this concern is misplaced. The infirmity does not generalize, for the laws of nature are enforced by higher authority.

FIVE

Schrödinger's Cat

Reality is nothing but a collective hunch.

Lily Tomlin

QUANTUM MECHANICS IS THE DETERMINISTIC LAW OF motion of very small things—atoms, molecules, and the sub-atomic particles of which they are made.[1] It was discovered in the 1920s by physicists trying to reconcile numerous strange and highly embarrassing experimental facts that seemed fundamentally incompatible with Newton's clockwork: the tendency of atomic vapors to emit light with distinct wavelengths, the tendency of hot bodies to glow with a color and intensity that increases with their temperature; and the laws of chemical bonding and radioactivity. The solution to the problem turned out not to be abandonment of the clockwork but a profound conceptual revision of its machinery. It is a beautiful case history of how science advances by making theories conform to facts rather than the other way around.

Learning quantum mechanics can resemble an out-of-body experience.[2] Things that cannot be become matter-of-fact truth, words acquire meanings that are the exact opposite of their customary ones, and commonsense reality gets turned on its head. Attending classes on the subject is like listening over and over again to Abbott and Costello's *Who's on First.*[3]

By far the craziest aspect of quantum mechanics is its mixture of Newtonian clockwork determinism and rather spooky probabilistic indeterminism, the latter invoked as needed depending on the experimental circumstances.[4] It is part of the lore of quantum mechanics that the act of measurement itself interrupts the deterministic time evolution—a kind of anthropic theory of reality not unlike Bishop Berkeley's famous proposition that a tree falling in the forest makes no sound.[5] This is absurd. A thing cannot be deterministic only when people are not looking at it. The probabilistic rule nonetheless describes certain experiments quite accurately and is in this sense true. How a certain rule could result in an uncertain experimental outcome is an important and interesting question.

The absurdity of the quantum observational paradox was deeply understood by Erwin Schrödinger, one of its inventors, who captured it with delicious irony in his famous thought experiment with a cat.[6] He imagined a closed box containing a cat, a single radioactive atom, a Geiger counter, and a cyanide capsule rigged to drop into a bucket of acid when the Geiger counter clicks.[7] The function of this contraption is to kill the cat with certainty if the atom decays. The deterministic rules of quantum mechanics then say that a mysterious quantity called the wave function leaks out of the atom slowly, the way air escapes from a balloon, so that a finite but constantly diminishing amount of this wave function is still inside. However, the physical meaning of the amount left inside the atom is a probability that the atom has not decayed *when one measures* it, that is, when one opens the box to see whether the cat is still meowing. Until the measurement is performed, however, the system is inherently a combination of alive and dead cat. The ludicrousness of this idea is self-evident, especially to anyone who has encountered an actual dead cat. Schrödinger intended it thus.

The ludicrousness of this idea is self-evident.

Zany illogic of this kind is almost always symptomatic of a missing idea. The Abbott and Costello routine is based firmly on this principle, as is the entire wacky world of Gracie Allen: "I know Babe Ruth has a twin brother because I read that his double won a game for the Yankees." "What does he call himself?" "Oh, you're so silly. He doesn't have to call himself. He knows who he is."[8]

The missing idea in the case of quantum measurement is emergence, specifically the principle of symmetry-breaking required for the apparatus to make sense.

The history of the quantum measurement paradox is fascinating. There is still no general agreement on the matter even after eighty years of heated debate. For some physicists, such as myself, the emergent nature of measurement is obvious and something responsible professionals do not waste time discussing. For others it is unspeakable heresy. The reason for this disagreement is that the arguments are subtle and not transparently resolved by experiments one can presently do. Scientists have ideological positions just like everyone else, especially in conflicted situations, and sometimes the consequences are bizarre. The Schrödinger cat has grown over time to become a symbol of transcendence, a meaning exactly opposite to the one Schrödinger himself intended. It has acquired quasi-religious overtones, so that twisting one's mind around to understand this cat is often viewed by students as a step on the path to enlightenment. Unfortunately, it is not. In science one becomes enlightened not by discovering ways to believe things that make no sense but by identifying things that one does not understand and doing experiments to clarify them.

The thing one does not understand in the case of the cat paradox is the measurement process itself. This quickly becomes apparent when one attempts to describe the measurement apparatus quantum-mechanically. In every case of ostensible indeterminism, this turns out to be impossibly difficult because the number of atoms is too large. In the case of the cat, for example, measurement might entail removing the top of the box and shining in a flashlight, or even leaving the top on and just sniffing. The impracticality of being tested against simpler explanations is something quantum indeterminism has in common with fantastic theories of the pyramids or arguments that extraterrestrials must now be running our government. There remain a few logical loose ends. Moreover, close inspection reveals that the number of atoms is *necessarily* too large, for the apparatus would not work if it were small. Detecting the radioactive decay of an atom using another atom, for example, makes no sense, since it would amount to substituting one tiny unmeasurable thing for another. But measuring with a tube of gas connected to a high voltage supply and an amplifier—a Geiger counter—makes perfect sense. Evidently there is something about the human concept of "measurement" that requires an apparatus to be large.

Once we recognize that largeness is a key factor, the mystery is not hard to resolve: All quantum detectors are made of solids, and thus all of them exploit the symmetry-breaking characteristic of the solid state, an effect that occurs only in the limit of large size. To qualify as an observation by the conventional human definition, a thing must not be changed by the act of observing it. An example of something *not* qualifying as an observation is my asking my neighbor his opinion on whether his department chair is having an affair with the previous chair's wife. I will get a different answer depending on whom he thinks I will talk to, and moreover, the answer may change from one day to the next as the winds of intrigue blow about. The only way I will get a consistent observation is if the various members of his department communicate with each other, hash the matter out, and decide collectively what the story is. We commonly speak of opinion "crystallizing" on subjects such as this. The physical version of this effect is that the various delicate quantum parts of the experiment cooperate to become a classical object obeying Newton's laws. When you read the meter on a Geiger counter, for example, you know with certainty that the value will be the same when you reread it an instant later, because the needle is a cumbersome, solid object. If I hear a click coming out of the speaker, the student across the room will hear the same click a fraction of a second later with one hundred percent certainty—unless he was not paying attention, in which case he will be history soon. But at the level of the atomic disintegration itself this is not true, for the system in question is easily disrupted by the act of observation. The apparatus works by transforming a quantum signal to a classical one by means of the emergence of objects.

One reason symmetry-breaking is so difficult to deduce from the underlying laws of quantum mechanics is that the world is configurationally entangled. Entanglement is a colorful term that brings to mind knotted electrical cords and unhappy experiences with discount fishing reels, but it is actually more like income tax. Recall that in an income tax calculation the final outcome is one simple number—the amount of tax you have to pay—but there are complex, interdependent rules on the way. Thus to your total wages, tips, etc., please add taxable interest from schedule B, except for tax-exempt interest, but write it here anyway, then add business income from schedule C, capital gain from schedule D, and many other things like this, then subtract moving expenses, after first checking form 3903 that probably prevents you from making this subtraction, then subtract itemized deductions, which include state income tax and mortgage interest, except if you made too much money, in which case add a fraction back that depends on details, and throw in some job expenses unless you have a job, then reckon your tax from the total in one of three ways that are all equivalent, then see page 34 for the alternative minimum tax that we forgot to tell you about, then write a big check. The wave function of a quantum system is like this. It is a rule by which the various inputs—in this case particle positions and orientations rather than income and job information—are converted to a number. The state of a quantum system, like the state of the tax system, is defined at any moment by this rule. Deterministic motion in quantum mechanics means logical and systematic evolution of the rule as time advances. Entanglement means interdependency in the rule. The entanglement of quantum mechanics is, however, vastly worse than that of income tax because everything is correlated with everything else. An apt tax analogy would be a rule for reckoning the total revenue to the government in which Joe's deduction depends on how much Alice spent on Caesar salad and whether George got a new truck. Expand this from the number of taxpayers to the number of grains of sand on all the beaches in the world and you have an idea of the problem of quantum entanglement in a small body such as a sugar cube.[9]

Quantum entanglement is one of those things that is easy to understand but almost impossible to believe—like free checking or protestations of innocence from tobacco executives. Nonetheless it is true. The simplest and most direct of the many experiments verifying its validity is atomic spectroscopy. Atomic vapors emit very specific wavelengths of light, whose exact values depend on the atom but whose sharpness and distinctiveness do not. The wavelengths are accounted for with enormous accuracy by the rules of motion of entangled electron wave functions. Moreover, these rules comply accurately with a property of this light known as the Ritz combination principle, which requires the observed frequencies always to be differences of more fundamental ones. There is much concern lately in demonstrating the entangled nature of quantum mechanics, but in truth it is demonstrated every day with great precision by the light emitted from atoms.[10]

Entanglement is hard to believe, in part, because the very emergent phenomena that enable us to control it also make it hard to see. If the freight train is coming, we need not consider its correlations with nearby insects to know that stepping off the tracks would be a good idea. It is also impractical to measure the mass of the unfortunate insects by carefully observing the train's jerks as it hits them, even though this is possible in principle. The insects have become effectively unobservable. It is similarly difficult to detect the effects of quantum entanglement in the motion of a voltmeter or the click of a loudspeaker. This is not simply a side effect of building the detector out of solids, however, but the actual detection strategy itself. The apparatus works like the train. The quantum entanglement in it has not disappeared but has simply ceased to have experimental consequences that matter.

The probabilistic nature of quantum measurement arises not from magic but from the working of amplifiers, the bridges between the quantum world and the classical one.[11] A simple prototype for such an amplifier is a bowling ball poised in a shallow dimple at the top of a hill.[12] This ball is a sensitive detector of forces, for once nudged out of the dimple in a particular direction it will accelerate down the hill in that direction until it reaches the bottom going at great speed. The shallower the dimple, the more sensitive the detector. In the limit at which the dimple disappears altogether, the ball becomes infinitely sensitive and capable of detecting forces that are arbitrarily small, including quantum forces, such as the recoil of an atomic decay. But this highly idealized quantum mechanics problem is sufficiently tractable that it can be solved in its entirety, atom plus detector, without postulating indeterminacy. One finds that the arrival of the ball at the bottom of the hill, where it is good and classical, is predicted with certainty by its arrival halfway down, just as Newton's laws require, and that both are correlated with the decay of the atom, but that the moment of arrival is uncertain. This occurs because the entire concept of "arrival" is emergent. So is the death of Schrödinger's cat, to which this example is aptly analogous.

The emergent nature of the principle exploited by quantum amplifiers causes them to have certain universal properties, notably the tendency to make false alarms. The ball on the hill is only approximately Newtonian, and will demonstrate this by rolling off of its own accord, no matter how precisely it is positioned at the top, if one waits long enough. This is nicely captured by the famous quantum mechanics problem in which one is asked to calculate the time a pencil can be made to balance on its point. The answer is about five seconds. For a real pencil it is even less because of thermal disturbance and wind, but five seconds is the fundamental limit. It is very generally the case that more sensitive amplifiers generate more quantum noise (the technical term for such mistakes) and that there is a fundamental relationship between sensitivity and noisiness. This is usually expressed abstractly as a Heisenberg uncertainty paradox, but it amounts to a pencil stood on end.

The generation of uncertainty by amplifiers resembles the generation of vacuousness by news organizations when there is no news. In politics things are often not "real" until they are widely discussed, so news media effectively make small events real by amplifying them. If the reported event is already fairly large, such as a troop movement or a cut in the discount rate, the amplified version is a reasonably faithful copy of the original. But if the event is small, e.g., a pork-barrel amendment or an unintentional but inflammatory misstatement, the amplified version can vary significantly from one report to the next and in this sense become uncertain. The limit of this process is reached when there is nothing at all to report, at which point the reporters begin interviewing each other and filling up air time with each other's opinions. Thus we have Paula Zahn asking Wolf Blitzer what he thinks the President's position will be on the upcoming tax cut fight, and so forth. In the news business this is called a slow day. In physics it is called quantum noise.

The emergence of conventional physical reality out of quantum mechanics is harder to grasp than the emergence of political structures out of news, however, because the starting point is so otherworldly. Quantum-mechanical matter consists of waves of nothing. This is a tough concept, so one traditionally eases students into it by first explaining something called the wave–particle duality—the idea that particles are Newtonian objects that sometimes interfere, diffract, and so forth, as though they were waves. This is not true, but teaching it this way prevents the students' mental circuits from frying. In fact, there is no such duality. The entire Newtonian idea of a position and velocity characterizing an object is incorrect and must be supplanted by something we call a wave function, an abstraction modeled on the slight pressure variations in the air that occur when sound passes. This inevitably raises the question of what is waving—a wonderful instance of the trouble one can create by using an ordinary word to describe an extraordinary thing. In customary usage a wave is a collective motion of something, such as the surface of the sea or a bleacher full of enthusiastic sports fans.[13] It makes no sense for a conventional wave to exist outside the context of something doing the waving. But physics maintains a time-honored tradition of making no distinction between unobservable things and nonexistent ones. Thus even though light behaves as though it were waves of some substance—referred to in the early days of electromagnetism as ether—there is no direct evidence for this substance, so we declare it to be nonexistent. For similar reasons we accept as nonexistent the medium that moves when waves of quantum mechanics propagate. This is a problem considerably more troublesome than that of light, however, because quantum waves *are* matter and, moreover, have measurable aspects fundamentally incompatible with vibrations of a substance. They are something else, a thing apart. The analogy I like best is Christina Rossetti's:[14]

> *Who has seen the wind?*
> *Neither you nor I:*
> *But when the trees bow down their heads*
> *The wind is passing by.*

Unhappily, the otherworldliness of quantum mechanics is a convenient justification for indulging in even more otherworldly "interpretations" of it that miss the forest for the trees.[15] The convoluted nature of these arguments infatuates the undergraduates but annoys the rest of us because they boil down in the end to attempts to describe quantum mechanics in terms of behavior that emerges from it, rather than the other way around. They are, in other words, symptoms of a failed worldview. One tries to be nice about this, but the temptations to be mean are sometimes irresistible.

One of the lessons we learn as we age is that misperceptions can appear to cause craziness where there actually is none. This is the source of much good humor, the universal appeal of which comes from the universality of the experience itself. The joke works particularly well if the protagonist is in deep denial of some essential thing. Early in my graduate student days I lived in a seedy apartment that I shared with several other students, who rotated in and out as professional constraints dictated. For a brief period one of these roommates was a warmhearted fellow from Cameroon who was studying engineering. He was an impressive person, particularly for his verbal ability, for English was his second foreign language after French. He had an interesting family too, including a cousin who was a recording star for Decca records. This cousin and a buddy once flew over from Paris to stay with us for a few weeks, so I got to hear his record. I did not like it very much. It was French disco, which would go tika tika tika tika for a long time and then pause long enough for him to say "ugh" and then continue with more tika. On this particular occasion they brought lots of presents with them, including food. Now, unfortunately for them, our place was terribly infested with cockroaches. It was not possible to eradicate the little monsters, even though we had complained bitterly to the landlord and tried several times to do it ourselves. They would just scurry next door to wait out the attack and then reoccupy after it was over. I do not know where they lived or what they ate, but they clearly loved replicating and having gigantic parties in the kitchen after the lights were out. We would find them in daylight in the darnedest places, such as behind a matchbook, inside the stove top, or underneath the forks in the silverware drawer. Like most people in this situation, we took to washing everything assiduously before we ate and keeping all open food containers, such as cereal boxes, in the refrigerator. You can thus imagine my surprise when I came home from work the day these guys blew in from Paris, reached in the cupboard for the peanut butter, and came face to face with the carcass of a dried animal about the size of a rabbit. In horror I called my roommate into the kitchen to ask him about it. "Messi, you can't keep a rabbit in here," I said. He looked at me without comprehension for a moment and then smiled broadly as he finally understood. "Ho ho ho," he said, "There is no problem. That is not a rabbit."

The Quantum Computer

Ours is the age proud of machines that think and suspicious of men who try to.

H. Mumford Jones

D RIVING TO WORK ONE MORNING I HEARD A FASCINATING allegation on the radio that women understand computers better than men do.[1] The speaker did this only indirectly and was careful to be politically correct, but the point of her remark was nonetheless clear. After she explained her position I could see that she was probably right. Men always want to tinker with the computer, she said, take it apart, add memory and peripherals, and so forth, while the women concentrate on more important things like sending out hundreds of email invitations to a wedding shower. This is fully consistent with my own experiences with technological things. When our car breaks down I obsess on figuring out what happened while my wife just wants to spend a lot of money to get it fixed and go to the movies. Women just seem to intuitively understand better than men that how a thing works matters much less than what one uses it for.

Computation is based on an enormous tower of functionalities.

Computers are an especially helpful instance of this technological fact of life because they are so transparently hierarchical. At the highest level they are tools that store and process email, manipulate more formal written communications, and allow one to search for deals at Internet auctions. (There are less practical uses, such as tasteless video games, secret pornography downloads, and trading copyrighted songs and movies, but these waste time and do not count.) At the next level down one has the processor, motherboard, and expansion slots containing wonderful things with names like voodoo and rage, so powerful that they require extra fans. Below this one finds silicon microchips with their fabulous webs of wires and diffused transistors, and below this the orderly lattice of silicon atoms through which electrons and holes propagate.[2] It is possible to send out all those shower invitations without thinking about it too carefully because of the reliability of an enormous tower of functionalities, each resting on the one below and supporting the one above. How each level works is immaterial. The invitations could just as easily have been sent out by little gnomes with pads of paper and miniature telephones, although they would probably have demanded more money.

Computers are machines. Like any other machine, such as a lawn mower or steam engine, a computer works by moving matter from place to place. Because the matter in question is composed of electrons solely, it can be made to move easily and with blazing speed, but it is still conceptually the same as a piston rod or crankshaft in a car.[3] In the end, the objective of computer engineering is still to get an assemblage of mechanical linkages to cause some physical thing to occur, such as deposition of ink on a page, motion of a loudspeaker cone, or twisting of the liquid crystal in a display pixel. Computers are often touted as the magical technology of the twenty-first century, but they are actually the crowning achievement of the nineteenth.

A key difference between computers and other kinds of machines is the ease with which their mechanical linkages can be modified. The modification process is called programming, and it has the genteel appearance of a term paper being typed, except that there is more coffee consumed and more swearing.[4] But looks can be deceiving. This activity is not term paper composition at all but auto shop. It involves construction of complex mechanical relationships between simple parts that then either function or don't depending on the workmanship. One has simply traded in lathes and torque wrenches for pencils and keyboards.

There are several qualitative differences between computers and cars brought about by the ease of modification. For example, the economics of engineering is fundamentally altered by driving the cost of making the physical microchip way below the labor cost of programming. This is why software costs so much, and why its monopolization is so different from that of steel, railroads, or oil.[5] Programming is also sufficiently similar to day-to-day use of the computer that the two become mixed up in people's minds as a kind of super abstraction of thinking. In the world of computers one begins to confuse play with work, work with play, and business activity with fundamental meaning. Computation, as most people experience it, is separated by complex layers of economic activity from the basics of the machines themselves and is in this sense a classic case of emergence. Modern computer programs are constructed by enormous teams of people, each of whom understands only a small fraction of the task, and these programs often wind up interacting with each other in ways their creators could not have imagined. This sociological phenomenon is a logical implication of the simple fact of cheap programmability made possible by the agency of electricity.

The trick to making modification easy is eradicating the difference between cause and consequence using transistor action. This has a simple analogy in one's own thought. I will remove my hand quickly from the stove top if the burner turns on, but I will also remove my hand if I remember a phone call I have to make. The complex circuitry that moves my hand can be actuated either by an external stimulus, such as fire, or an internal one, such as a surfacing memory. There is no difference between the two other than an abstract categorization. This can go awry in mental illness, in which case a person begins to confuse real events with imaginary ones. The transistor is an equalizer. It senses a motion of electrons in one wire and generates motion of electrons in another that is always the same size, regardless of how small the first motion was. This causes the motions in computers to be all-or-nothing affairs, with each wire being in either an on or off state and never in between. It also causes the measurement of a given wire to contain no information about where the signal originally came from. The decision to be on or off could have been based on an external stimulus, another transistor, or a huge nested cascade of transistors. There is no difference.

The signals in computers are Newtonian. We sometimes lose sight of this fact, since computers tend to be viewed as mysterious in the same way that quantum mechanics is, but this is exactly backward. The mysteriousness of computers comes from the emergent nature of their functionalities, not from microscopic considerations. At the level of the transistors themselves computers are grounded firmly in the idea of absolute certainty in measurement, for only this is compatible with the idea of on or off—right or wrong—at any given moment. Not only are transistors Newtonian concepts, they *create* Newtonianness by outputting large motions of electrons in all circumstances. In the process of doing this they generate heat—lots of it.[6] This is why modern processor chips are hot to the touch, and why they will die if their dedicated fan malfunctions. The generation of heat is fundamental to maintaining reliability. To see how this could be so, it is helpful to return to the famous example of a pencil on its tip. In practice, the decision to fall left rather than right is permanent, because the pencil dissipates all its energy into heat when it crashes down on the table. If it didn't—if the encounter with the table resulted in a perfectly elastic bounce—the pencil would right itself again and make the left–right decision a second time, perhaps with the opposite outcome. So the dissipation of power and the generation of heat are essential to decision-making, particularly in situations involving initial delicate balance, and thus to the functioning of all modern computers. (We might say the same of human institutions such as companies and government: the decisions that count are irreversible.)

Two small modifications of the transistor design allow one to build actual computers. One involves giving the transistor two input wires and causing it to be on if either of the two inputs is on. The other involves a negation, so that the transistor is on if the input wire is off and vice versa. These two design elements are called *logic*, and they form the conceptual basis of all computer circuitry. Modern computers are simply an enormous network of logic and a clock—a small bit of circuitry that switches a wire on and off in a regular way, like a heartbeat. The clock hearts of modern home computers beat very fast—effectively about a billion times a second—but sturdily and valiantly. I have had a couple of computers die of heart failure, but such deaths are rare. Computers nearly always become obsolete long before the grim reaper comes to call.

There is a great deal of interest lately in the quantum computer, a fundamentally new kind of computational hardware that would exploit the entanglement of the quantum wave function to perform calculations presently impossible with conventional computers.[7] The most important of these is the generation of enormous prime numbers and the quick factorization of other enormous numbers. The impossibility of factoring a number that is the product of two large primes in reasonable time with conventional computers is the basis of modern cryptography.[8] However, quantum computation has a terrible Achilles heel that becomes clear when one confronts the problem of reading out the answer: the effects that distinguish quantum computers from conventional ones also cause quantum indeterminism. Quantum-mechanical wave functions do indeed evolve deterministically, but the process of turning them into signals people can read generates errors. Computers that make mistakes are not very useful, so the design issue in quantum computation that counts is overcoming mistakes of measurement. A textbook method for doing this is to place a million copies of the same experiment in a small box and measure something they do collectively—generate oscillating magnetic fields, for example, as occurs in a quantum computer built with electron spins. The damage inflicted by the measurement process then affects only a few copies, leaving the rest intact. This trick is so powerful that variations of it enable you to read out the entire wave function of any quantum computer, at least in principle. However, a logical implication of this ability is that you have created not a fabulous new kind of digital computer but a conventional analogue computer—a type of machine we do not use in the modern era because it is so easily disrupted by noise.[9] Thus the frenzy over quantum computing misses the key point that the physical basis of computational reliability is emergent Newtonianness. One can imagine doing a computation without exploiting these principles, just as one can imagine proving by brute force that broken symmetry occurs, but a much more likely outcome is that eliminating computational mistakes will prove to be fundamentally impossible because its physical basis is absent. The view that this problem is trivial is a fantasy spun out of reductionist beliefs. Naturally, I hope I am wrong, and I wish those who invest in quantum computing the best of luck. I also ask that anyone anxious to invest in a bridge in lower Manhattan contact me right away, for there will be discounts for a limited time only.

The *real* quantum computer, of course, is good old silicon.[10] The principles of semiconduction on which transistors are based, and the difference between conventional conducting wires and insulators, are highly quantum-mechanical. This fact was not apparent when semiconduction was discovered by Ferdinand Braun, who stumbled upon it in a number of metallic sulfides, notably the lead ore galena, in 1874.[11] Only much later, in conjunction with the development of radar and the related invention of the transistor, was a systematic understanding of the quantum nature of these effects worked out, mostly by the legendary John Bardeen. Crystalline insulators conduct electricity poorly because all of their electrons are tied up in chemical bonds. In the specific case of silicon, for example, each atom has four neighbors and four electrons available for bonding—a number exactly exhausted by the usual rule of two electrons per bond. In contrast to extremely good insulators such as quartz or table salt, however, the chemical bonds of silicon are weak and easily disrupted. Once ripped out of its bond, an electron is free to move about in the silicon, as is the hole left behind. The rectifying and amplification actions of semiconductor devices all come from manipulation of these freed electrons and holes by means of chemical modifications and attached wires. The quantum mechanics that matters regulates the bonding rules and the motion of the freed electrons and holes.

Electrons and holes move through cold crystalline silicon as if it were not there.[12] This astonishing fact is central to the working of transistors, and it is why efficient ones can never be made from non-crystalline substances such as rubber or plastic.[13] Indeed, the key technical breakthrough that ushered in the silicon age was not the invention of the transistor but the invention of zone refining, a method of systematically eliminating chemical and structural imperfections of crystals. The ability of electrons and holes to move ballistically through the lattice is not obvious at all, for a piece of silicon is conceptually no different from a giant molecule and must therefore be characterized by the highly entangled motions of *all* the electrons, including those in the bonds. The resolution of this problem is that the entanglement is rendered irrelevant by emergence. It turns out to be exactly and universally the case that crystalline insulators have specific collective motions of all the electrons that look and act as though they were motions of isolated electrons. The only effect of all their awful underlying complexity is to make the acceleration mass slightly different from that of a free electron and to effectively reduce the strength of the electric forces. The electric charge of a hole is obviously opposite to that of an electron, since it represents an electron deficit. An engineer speaking of an electron or hole is really speaking about one of these complex collective motions, not an isolated particle. For engineering purposes this complexity does not matter any more than it matters how computers send out shower invitations. The important thing is that the particle-like nature of the collective motion is exact and reliable.

Electrons and holes in silicon are magnificently quantum-mechanical. Despite being not free at all but horrendously entangled, these objects provide some of the most accurate tests of quantum mechanics that have ever been obtained. A beautiful example is the line spectroscopy of phosphorous impurities. Phosphorous atoms added in small amounts to melted silicon substitute for silicon atoms in the lattice when it crystallizes. The substituted phosphorous atom uses up four of its five outer electrons making chemical bonds and gives up the fifth to wander about. When the temperature is reduced to extremely low values, however, the errant electron finds its way back to the site and binds there, just as an electron would bind onto a proton to make a hydrogen atom.[14] However, rather than emitting visible light at distinct wavelengths, the electron bound to the phosphorous impurity emits infrared light at distinct wavelengths because the electric forces binding it to the phosphorous site are powerfully mitigated. This light can be detected with conventional infrared spectrometers. Not only is the impurity spectrum analogous to that of an atom, it is *physically indistinguishable* from that of an atom, except for the specific wavelengths of light emitted. The collective nature of the object doing the binding has all but disappeared. There are lots of experiments like this, for inside a piece of silicon is a miniature world in which the forces of electricity are reduced, the masses of electrons are changed, and the electron has a sibling of the opposite charge with which it can annihilate to make light.

The quantum nature of electrons and holes almost certainly imposes a fundamental limitation on Moore's law, the celebrated observation by Intel founder Gordon Moore that the number of transistors on a given area of silicon tends to double every 18 months.[15] Moore's law is one of the main reasons computers have continued to surprise us when the basic principles underneath them are so simple. Back in the beginning of the computer age it was discovered that making transistors and wires smaller so as to pack more of them onto a silicon chip also made them more reliable. Thus began the race to achieve higher and higher densities that continues to this day. Right now, chip manufacturers are fighting terrible problems of heat generation and optical lithography size constraints unlike anything they have faced before, but nearly everyone believes these can be overcome in time to preserve Moore's law down to the quantum limit. In about a decade or so, however, the transistors will get so small that they will become quantum-mechanical—and thus begin making mistakes. When this occurs it will mark the end of a remarkable time in history in which the implications of a small physical discovery exploded into the economy and changed the world.

One of the more interesting trends of the computer age is that physical science students are increasingly unwilling or unable to write computer code. I was very upset when I first observed this and took stern measures in my department to counteract it, much to the students' chagrin, for I myself am very good at coding and consider it something any self-respecting technologist should know how to do. Eventually, however, I realized that the students were right and I was wrong, and stopped the crusade. Computer programming is one of those things in life, like fixing one's own car, that is fascinating, fun, useful—and unacceptably time-consuming. The truth is that it is no longer cost-effective for most well-educated people to program their own computers, or even to learn how to do so. The wise use of time is to spend a few bucks to buy a program that does what one wants or, in extreme cases, search the internet for free software.

When I was a graduate student, in the early 1970s, the economics were exactly reversed. Student labor was cheap and computers were hideously expensive mainframes that occupied entire floors of university computer centers. They were pampered affairs, with legions of attendants working in shifts around the clock and special air conditioning with power backups. We wrote computer programs for these behemoths late at night on grey metal machines about the size of bears. One of these machines would hum away with its motor running until one struck a key, at which point it would tremble a bit, go chunkh, and put a crisp new hole in the card one was punching. After one was done with a card, one would hit the feed key, and the machine would rotate the card klicka-ka-chunkha-chunkh to the bottom of the growing stack and feed in a new blank one. The computer programs we wrote were realized as decks of cards punched in this way, held together with rubber bands and stored in cardboard boxes. Running the program involved submitting one's deck to an attendant, who would feed it into a card reader, an apparatus that looked like a gasoline-powered wood plane and sounded rather like a vacuum cleaner with a leaf caught in its fan. The printer would all the time be whining away in the background under its metallic sound hood and frantically throwing off page after enormous page of white computer paper as though it had gone berserk. Every few minutes an attendant would retrieve this output, which would require opening the sound hood briefly, thereby filling the room with unbearable screeching. The attendant would then tear the output at appropriate seams and stuff it into bins for pickup by students. This output consisted mostly of incomprehensible operating system mumbo-jumbo with the stuff one had actually calculated on the last page—unless the code had a mistake, in which case one would get either a thin nothing or an inch of meaningless core dump output, depending on the severity of the mistake. The expense of all this was incomprehensible. I remember talking with one of my fellow students just after he had submitted a deck in three boxes and seeing his hands shake. Ah, those were the days.

The most famous deck story of all time is the box containing an enormous hydrodynamic simulation code that somebody dropped, causing cards to fly everywhere. The program in question was promptly christened Nixon because it would clearly never run again. But happily it did run again and became the nucleus of the classified program Lasnex, the current workhorse of laser fusion simulation.[16]

The joke about gender bias in computer skills thus has at its core the more important observation that we owe the existence, reliability, and utility of computers to principles of organization—including economic ones. That women have an easier time understanding the supremacy of organizations than men is not news, for this was known to the ancients and recorded in numerous places, notably the *I Ching*.[17] According to Taoist philosophy, the universe is impelled forward by the conflict of two opposing principles, yin and yang, that constantly produce and supplant each other. Yang represents maleness, the sun, heat, light, dominance, and so forth. Yin represents femaleness, the moon, material forms, cold, submission, and so forth. Yang, the sunny southern side of the mountain, creates, while yin, the shady northern side, completes the created thing. One might say that we are presently in an age of yin, and even though computers were brought into existence by yang, they have reached their full potential only under the dominance of yin. A more direct western way of saying this same thing is that computers were originally conceived as dogs but now have become cats. The machine one brings home from the store is clever, self-serving, constantly underfoot, and always scheming how to get you to do what it wants. But when you lobotomize the thing, strip away its sophistication, and reach down past the facade to the wires, transistors, and algorithms underneath you find unquestioning obedience, steadfast loyalty, straightforwardness, and simplicity—i.e., a dog.

SEVEN

Vin Klitzing

If scientific reasoning were limited to the logical process of arithmetic, we should not get very far in our understanding of the physical world. One might as well attempt to grasp the game of poker entirely by use of the mathematics of probability.

Vannevar Bush

It is difficult to keep professional concerns in focus while floating down a river on a tour boat on a breezy summer afternoon. Scientists are always complaining about the agony of writing all those proposals, delivering all those technical presentations, and accruing all those frequent flyer miles, but such complaints are disingenuous, and they are exposed by such moments as fraud. Most of us are willing to pay the price for the perquisites, and grumble in public only to prevent other people from discovering how pleasant our lives actually are. Right now, the inevitable quid pro quo is far away, for it is warm and there are fields and orchards passing by. Science is a tough job, but somebody has to do it.

I am on the Neckar as guest of a group of alumni of the Max Planck Institute in Stuttgart, who have hired the boat as a sixtieth birthday present for the legendary Klaus von Klitzing.[1] It is a friendly bunch of people, many of whom I have known since the early 1980s

71

when I first began to write theoretical papers about the von Klitzing effect. Most are locals, but a few, like myself, are from abroad. In the mix of nationalities, Japan and America are especially well represented, as expected of a community of semiconductor physicists, but so are Israel and Russia, with smaller contingents from England, Brazil, and Mexico. Everyone is here for the common purpose of creating a memory for Klaus, a citizen of the world and an inhabitant of what we must now, apparently, call "Old Europe."[2]

Ageless and enthusiastic as always, Klaus is unaware of the real surprise coming downriver—a small vineyard on a bluff rented for him by his friends. He chats away happily as the boat rounds the bend, and then stops abruptly as he spots the big sign with his name on it on the hill. It is manned by a couple of students who drove up in the morning to set it in place. They see that they have been discovered and wave. Klaus instantly figures out what has happened and becomes quite animated, but it is too late. Anticipating this moment, the conspirators have secretly distributed glasses of champagne, with which the birthday is now exuberantly toasted, causing the boat to rock a bit. Klaus is speechless. The boat comes to a brief halt in front of the vineyard for a few photographs and testimonials, including solemn promises to press the wine without cheating by mixing with inferior grapes. It is to be privately bottled and distributed under the label *Vin Klitzing*.

While the key surprise recedes upstream, others are still in store. The boat docks at the medieval town of Besigheim, where it is met by a welcoming committee of curious locals, a terrific high school band, and three gentlemen dressed in eighteenth-century livery. One of these, evidently the leader, sports a three-foot-high stovepipe hat with a wide brim and carries a monstrous glass of wine. He delivers his boatload of guests a ceremonial invitation to enter the town, then leads them through the cobbled streets to a great banquet hall laid out in preparation for a feast. The students are overjoyed to discover it is an all-you-can-eat affair, no doubt in deference to them. The stuffed guests are then led from the hall on a tour of the medieval wall surrounding the town and reinforcing the protection of the converging rivers below, which bubble and sparkle in the sun as they have for centuries. The party then reboards the boat and climbs back home through mossy locks in the failing light, singing and sampling wine after excellent wine from the stores, for this trip is a tasting tour in addition to everything else.

The extremes to which these people have gone to honor Klaus reflect the esteem in which he is held. Part of this enthusiasm is admittedly a local phenomenon. On the day in 1983 when his Nobel Prize was announced, for example, they interrupted daytime television in Germany, something utterly unthinkable in America, for continuous coverage of the event. He was only the fourth physics Nobel in Germany since the Second World War, a particularly sensitive matter, since modern physics was invented in Germany in the first years of the twentieth century.[3] But he is lavishly fêted elsewhere in the world as well, especially in Asia, and is seemingly always on his way to and from honorary presentations and meetings in the far corners of the earth.

What he did to deserve this celebrity was to discover something that should not have been—a shocking reminder that human understanding of the world is finite, that our prejudices are not laws, and that quantum physics is magical or often seems to be.[4] He made his discovery in 1980 at the high magnetic field laboratory in Grenoble, where he was performing interesting but rather routine experiments on state-of-the-art electronic components. These were built to tolerances more exacting than those used in the microcircuit industry even today and cooled to ultralow temperatures for the purpose of enhancing any new effects that might be exploited in the next generation of electronics. He began thinking about an effect in these samples that had been seen before in which one of the measurements became abnormally steady over a range of magnetic field strengths. Motivated by curiosity, academic training, or just plain inspiration he resolved to find out exactly how steady it was by accurately calibrating the experiment. He discovered, to his amazement, that it was the same from one field strength to the next, one day to the next, and one sample to the next to an accuracy of better than one part per million. Improvements in sample quality and refrigeration technology have since improved this reproducibility to one part in ten billion. To put this accuracy in perspective, it is like counting every man, woman, and child on the surface of the earth without missing a single one. The discovery of this unexpected, unpredicted constancy rocketed von Klitzing to international superstardom in science, where he has remained ever since.

The measurement itself is simple—once you know what to look for—and has been reproduced in thousands of laboratories around the world, so we are sure it is right. When a magnet is brought into the vicinity of any wire carrying electric current, a voltage develops at right angles to the current flow. This occurs because electrons moving in the conductor are deflected by the magnet, just as they would be in free space, and so pile up on one side of the wire until the reaction voltage they generate exactly compensates the magnetic deflection. This is called the Hall effect, named after Edwin H. Hall, the physicist who originally discovered it in 1879. It is normally reported as a resistance, computed by dividing the voltage thus generated by the current. At ordinary temperatures the Hall resistance measures the density of electrons in the wire, and is therefore important in semiconductor technology, which works by manipulating this density. At very low temperatures, however, quantum mechanics intervenes. A plot of the Hall resistance versus density is no longer a straight line, as it would be at room temperature, but one that has acquired wiggles. In the case of the special kind of semiconductors von Klitzing was studying—field effect transistors like those in computer chips—these wiggles evolve into a staircase with extremely flat steps as the temperature is lowered. The heights of these steps are the universal quantized values of the Hall resistance.

After establishing its universality, von Klitzing quickly realized that the quantum of Hall resistance thus defined was a combination of fundamental constants—the indivisible quantum of electric charge e, Planck's constant h, and the speed of light c—all of which we think of as building blocks of the universe.[5] This fact has the obvious implication that you can measure the building blocks with breathtaking accuracy without dealing with the building blocks directly. This is deeply important and deeply upsetting to most physicists. The more thoughtful of them find it impossible to believe until they study the numbers, and even then suspect something to be amiss. But nothing ever is. The experiments are plentiful, consistent, and unassailable. Moreover, the accuracy of the von Klitzing measurement appears to improve without bound as the temperature is lowered and the sample size is increased. For this reason it has become the accepted definition for this particular combination of fundamental constants.

The impact this discovery had on physics would be hard to overstate. I remember vividly the day my colleague Dan Tsui brought the von Klitzing paper into the Bell Labs tea room and, barely controlling his excitement, urged everyone to think about where this astonishing accuracy could have come from.[6] No one had an explanation. We all knew that von Klitzing's samples were imperfect, and we thus expected them to vary. In processing semiconductors there are always variabilities one cannot control, such as structural defects in the crystal lattice, randomly incorporated dopants, amorphous oxides at the surface, ragged edges left over from optical lithography, bits of metal strewn about on the surface by clunky soldering irons when wires are attached, and so forth. These are known to influence other electrical measurements, for the matter is important for microcircuitry and has thus been extensively studied. But this expectation turned out to be wrong. As a result of theoretical work done after the fact, including some of my own, we now understand that imperfection has exactly the opposite effect, namely to *cause* the perfection of the measurement—a dramatic reversal worthy of the finest Greek drama.[7] The quantum Hall effect is, in fact, a magnificent example of perfection emerging out of imperfection. The key clue that this is so is that the quantization accuracy—which is to say, the effect itself—disappears if the sample is too small. Collective phenomena are both common in nature and central to modern physical science, so the effect is in this sense neither unprecedented nor hard to understand. However, the extreme accuracy of the von Klitzing effect makes its collective nature undeniable, and therein lies its special significance.

Over the intervening years, as I have lived inside theoretical physics and become familiar with its ways and historical currents, I have come to understand the von Klitzing discovery to be a watershed event, a defining moment in which physical science stepped firmly out of the age of reductionism into the age of emergence. This shift is usually described in the popular press as the transition from the age of physics to the age of biology, but that is not quite right. What we are seeing is a transformation of worldview in which the objective of understanding nature by breaking it down into ever smaller parts is supplanted by the objective of understanding how nature organizes itself.

If the quantum Hall effect raised the curtain on the age of emergence, then the fractional quantum Hall discovery was its opening movement.[8] The experimental setup that revealed the fractional effect was exactly the same as for the original von Klitzing effect, but the meaning was different. While the extreme reproducibility of the quantum Hall effect had been unexpected, the broad-brush behavior had not. Indeed, von Klitzing's interest in the matter had been sparked by a theoretical paper by Tsuneya Ando, now a professor of physics at the Tokyo Institute of Technology, in which figures very similar to the experimental traces later discovered actually appear.[9] The fractional effect, in contrast, was unanticipated by any theory and not analogous to anything previously known in nature. Dan Tsui and Horst Störmer discovered it accidentally one night while looking for evidence of electron crystallization, which is what prevailing theories said should occur. Instead, they found a miniature version of the von Klitzing effect at a magnetic field strength that should have been too high and at a value exactly one-third the ostensibly minimum allowed value, which should have been impossible. Von Klitzing always says he could kick himself for not finding the fractional effect, but he is being unfair to himself, for it was simply a matter of sample quality. (Imperfections cannot hurt the quantization accuracy, but they can, unfortunately, destroy the effect entirely.) Momentous discoveries often hinge on slight technological advantages. Dan, Horst, and I shared the 1998 Nobel Prize for work on the fractional quantum Hall effect—they for discovering it and I for constructing its first mathematical description.[10] I did not think of this discovery as revolutionary at the time, for my discipline is filled with astonishing quantum-mechanical things that require new mathematics to describe, but I have changed my mind over the years. The extreme perfection of the effect places it in a different category from its predecessors in the same way perfection of the original quantum Hall effect did.

The fractional quantum Hall effect reveals that ostensibly indivisible quanta—in this case the electron charge e—can be broken into pieces through self-organization of phases. The fundamental things, in other words, are not necessarily fundamental. That such fractionalization could occur in principle had been known for decades, and there was even an experimental literature arguing that particulate objects carrying fractional charge were responsible for electric conduction in organic conductors called polyacetylenes.[11] However, all of the arguments in place at the time of the discovery had flaws. The theoretical models in which the effect could be demonstrated conclusively were all one-dimensional and thus impossible to realize exactly in the laboratory. The organic conductors in question were always plagued with chemistry problems that prevented their experimental properties from being reproducible. It was always possible to evade fractionalization issues by arguing that the experiments could be explained without them—something that is always true in emergent phenomena but that often misses the forest for the trees. But the fractional quantum Hall discovery stopped this obfuscation in its tracks by virtue of its exactness. It is not possible to account for exact things with approximate theories. The observation of accurately quantized fractional quantum Hall plateaus *proved* the existence of new phases of matter in which the elementary excitations—the particles—carried an exact fraction of e. The excitations of the state first discovered by Dan and Horst carried charge $e/3$, an especially intriguing result in light of the charge $e/3$ carried by quarks, the allegedly fundamental constituents of protons and neutrons. Since then an immense cascading tree of such phases has been discovered, each characterized by a different small-denominator fraction.[12]

Once a person reaches a certain plateau of fame it becomes difficult to think of anything to give him that he does not already have. I had the unenviable task of delivering a technical lecture at the von Klitzing birthday colloquium that preceded the Neckar boat ride. Since I had not worked in semiconductors for years and had much less to say about the subject than younger people in the thick of things, I was in danger of making a fool of myself. I decided in the end to talk about emergent physical law—the aspect of the von Klitzing discovery that counts—and to use the occasion to present Klaus a seedling. I brought two, actually, one for his house and one for the Institute campus, for one learns over years of technical life to have backups. I explained in my presentation that these were sequoias, the mightiest of

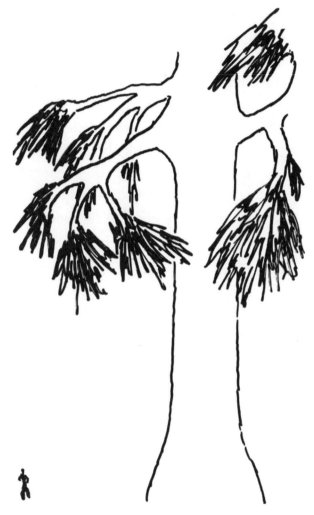

They appear to do well in the local climate.

trees, and native to the part of the world in which I grew up. I used to camp among them when I was a kid, although I did not understand until I left home how unusual they are. They are quite difficult to describe in plain words, for they are not vegetables so much as statements. I had been quite surprised to discover three fairly mature ones growing on Mainau Island in Lake Constance on my last visit to this part of the world.[13] They appear to do well in the local climate. This impression was confirmed by the people at the native plant nursery up in the Santa Cruz Mountains where I bought these little guys, who claimed that lots of people take them on airplanes, typically to Germany and Israel, interestingly enough. So I explained to the audience that the significance of the trees was less their pedigree than the fact that I brought them. I know that Klaus always flies economy class whenever possible, so he understood perfectly what it meant to sit ten cramped hours over the North Pole with a bag under one's feet. By the time all of us are dead, I said, these trees will be about the size of a conventional fir. By the time our children are old they will begin to look oddly large and a bit out of place. By the time seventy generations have passed—the time separating Julius Caesar from us—they will dwarf everything around them. There is no reason, with proper care, that they cannot live forever.

The important issue implicit in the von Klitzing discovery is not the existence of physical law but rather what physical law is, where it comes from, and what its implications are. From the reductionist standpoint, physical law is the motivating impulse of the universe. It does not come from anywhere and implies everything. From the emergentist perspective, physical law is a rule of collective behavior, it is a consequence of more primitive rules of behavior underneath (although it need not have been), and it gives one predictive power over a limited range of circumstances. Outside this range, it becomes irrelevant, supplanted by other rules that are either its children or its parents in a hierarchy of descent. Neither of these viewpoints can gain ascendancy over the other by means of facts, for both are fact-based and both are true in the traditional scientific sense of the term. The issue is more subtle—a matter of institutional judgment. To paraphrase George Orwell, all facts are equal, but some are more equal than others.

EIGHT

I Solved It
at Dinner

The subtlety of nature is greater many times over than the subtlety of the senses and understanding.

Sir Francis Bacon

BOB SCHRIEFFER, WHO SHARED THE NOBEL PRIZE IN PHYSICS for the theory of superconductivity, tells the following story. John Bardeen, his PhD thesis adviser and mentor, had just won his first Nobel Prize as an inventor of the transistor, and was on his way to Stockholm in December 1956 when the key ideas of the now-famous theory of superconductivity began falling into place. He found it maddening to be called away at this exciting time, but he had no choice but to go. When he returned in January, he and Bob began working day and night to hammer out the details of this theory and, in particular, to find ways of testing it by experiment. At a crucial moment in their work Mrs. Bardeen scheduled a dinner party. She invited one guest, a Swede, no doubt expecting that John would be sociable and entertain him. John was taciturn at the best of times, and was notorious for speaking slowly even when answering the simplest questions—presumably because he was thinking deeply about the possible answers to "How are you?" and all their implications. This was when he was *not* preoccupied with inventing a great new theory. So when the party rolled around it was, well, a night to remember. John barely said a word at dinner. He gave only the briefest answers to questions, asked none of his own, expressed no interest in his wife or the guest, and generally behaved like an incorrigible zombie. Mrs. Bardeen somehow got through the evening, packed off the guest, and began washing up when John wandered in with a strange smile on his face and remarked absent-mindedly, "I solved the heat capacity problem." When asked what in the world he was talking about, he said, "I did it at dinner."

This story always elicits warm smiles from people in my line of work because we remember learning that famous heat capacity formula as students and are aware that at least in some circles, John Bardeen is considered the greatest theoretical physicist who ever lived.[1] The fact that he did not fit popular stereotypes makes it all the more wonderful. John had none of the cult-figure status of Albert Einstein, nuclear weapons cachet of Robert Oppenheimer, or notorious intellectual arrogance of Wolfgang Pauli. He was instead a personable Midwesterner who quietly became the first person in history to win the Nobel Prize twice in the same field—the first time for the invention of the transistor and a second time for the theory of superconductivity.[2] Colleagues who began their careers in the 1960s have told me that John effectively invented the modern discipline of solid state theory.[3] He set its tone while working on the transistor at Bell Labs by painstakingly poring over experimental data, again and again confronting facts and trying to make sense of them with simple, transparent theories. When he failed in his initial attempts to make a field effect transistor, the device on which all modern microcircuitry depends, John focused his energies on understanding why. He correctly determined that the problem was surface states, an effect involving broken chemical bonds at solid surfaces, and urged Walter Brattain to try a different approach. The result was the point-contact transistor, announced in 1947,[4] the first primitive step toward the age of the microchip, which was finally ushered in only after the surface-state problem was overcome years later at Fairchild Semiconductor by means of a chemistry trick.[5] Nevertheless the invention of the original transistor set the standards of our discipline, and it still leads most of us to consider the highest achievement in science to be the rendering of facts down to their essentials so effectively that some practical invention becomes possible. This attitude comes directly from John Bardeen.

Bardeen may indirectly have owed his second Nobel Prize to William Shockley, his supervisor at Bell and winner, along with Brattain, of the transistor Nobel Prize. Shockley was a thorny character notoriously full of himself. When asked whether he donated to sperm banks that targeted scientists and other famous figures, for example, he is reported to have said, "Of *course* I gave," as if to refuse would be a disservice to humanity. A more typical response would have been to brush it off as a joke. He was known to consider scientists (as opposed to engineers) dilettantes and to despise them for their mental weakness. This warm and charitable attitude later blossomed into his famous theories of racial inferiority and eugenics.[6] Not surprisingly, he was enraged that physicists had actually invented the transistor while he, an engineer, had not. He took steps to muddy the waters of credit and to make the inventors' lives miserable. Brattain, who actually built the first transistor, refused to work for Shockley, and Bardeen left Bell on account of him. John emigrated to the University of Illinois, where he remained for the rest of his career. It was there that Cooper, Schrieffer, and he solved the superconductivity problem.

John Bardeen's scientific stature is so stupendous that it is hard for most of us to imagine him as a radical thinker with very human vulnerabilities. I recently heard a charming Bardeen story from Doug Scalapino, one of his close associates from the old days of superconductivity. Long after the transistor had been invented and the theory of superconductivity had been accepted, the two of them were playing golf in Santa Barbara. The politics of science came up, and John, who was having the usual trouble marketing some of his most recent work, remarked in an off-handed way, "You know, Doug, the Establishment is out to get me." Doug replied gently, "John, you *are* the Establishment."[7]

After the transistor, superconductivity was the right thing to worry about. The potential applications of superconductors are much different from those of semiconductors, but the central issue in each case is why some things conduct electricity and others do not. In a conductor some parts are free to move about while others remain rigidly locked in place—as though it were a cell phone with a complement of tiny loose screws inside that were free to bonk about when one turned it over. In the case of semiconductors, the loose screws can be thought of as caused by heating, since they disappear completely when the semiconductor is made cold. In a metal, in contrast, they remain present all the way down to absolute zero, and are thus inherently quantum-mechanical. Moreover, there are a lot of them. In a typical semiconductor like the ones in your personal computer or wristwatch, there is at most one carrier (an electron or hole) for every ten thousand atoms. In a metal there is one per atom. Where these objects come from, what absolves them of the need to make chemical bonds, and why they remain mobile even at the lowest temperatures are all profound questions. The matter central to all of them, as it turns out, is superconductivity.

The superconductor problem was difficult to solve, in part, because it required attacking an entrenched scientific shibboleth—the sea of independent electrons. In the early days of quantum mechanics it was discovered that many properties of real metals could be explained by postulating the electric forces between their electrons not to exist. It was not clear why this should work, but the properties of these highly idealized electrons were sufficiently simple that they could be calculated in a spare moment on a napkin, and these calculations matched experiment well. This fact turned out to be extremely useful in engineering and allowed one to anticipate reasonably accurately what metals should do in new situations. Unfortunately, it also had the effect of suggesting that matter *ought* to behave this way, a fundamentally incorrect idea. The forces between the electrons are, in fact, enormous, and their irrelevance to these experiments nothing short of astonishing. Metallic behavior is an emergent organizational phenomenon. The electron sea makes sense because the metallic phase has formed, not the other way around.

Bardeen, Cooper, and Schrieffer evaded the political problem of the electron sea by the brilliant tactic of making the superconducting state subservient to it. This is rather like the US Constitutional Convention making the President formally subservient to the Congress in declaring war. The maneuver assuaged the fears of the delegates about an imperial presidency and facilitated ratification of the Constitution, but in practice, the actual power to declare war was left in the hands of the President.[8] The superconducting state is, in fact, the parent of the conventional metallic state, not the other way around. But the theory reverses these roles by accepting the existence of the sea and accounting for superconductivity as a low-temperature subtlety caused by motions of the atomic cores around which the electrons flow. It would go away if the atoms did not move. But there is some fine print: The atoms in metals *always* move. The electron sea is *absolutely* unstable, meaning that it will become superconducting at sufficiently low temperatures if there is any atomic motion whatsoever. Thus the thing to which the superconducting state is ostensibly subservient is actually a mathematical fiction.

The key feature of the superconducting state predicted by the theory is something called the energy gap. The energy gap can be described with precise mathematics, but it is more helpful to think of the scene in Cecil B. DeMille's *The Ten Commandments* in which

Moses parts the Red Sea: the water draws back to make a channel bordered by cliffs, in defiance of the usual rules of flow, and the ground underneath becomes exposed and dry.[9] The channel thus formed facilitates the Israelites' escape from Egypt into the Sinai desert, and is measured by the miraculous conduction of Charlton Heston and the Israelites from one bank of the Red Sea to the other while the soldiers chasing them are blocked. In superconductors the gap is similarly measured by the miraculous conduction of electricity through a metallic sample while conventional motion of electrons is blocked. The experiment measuring the gap consists of two pieces of superconductor separated by a thin insulating layer—typically films of lead separated by a layer of lead oxide. A tiny electric current miraculously flows through this apparatus without any voltage applied across it at all—as though all impediments to electron motion had vanished—but large conventional currents flow only when the voltage exceeds a specific threshold value. This threshold is the energy gap. If the apparatus is warmed sufficiently to destroy the superconductivity, both of these strange behaviors disappear: the electron sea re-forms, the small supercurrent ceases, and the conventional current flows immediately when voltage is applied. The gapless electron sea is thus a high-temperature phenomenon that was mistakenly identified as elementary in early studies of metals because of inadequate refrigeration technology. In *The Ten Commandments*, the Red Sea was similarly misidentified by Pharaoh as a fundamental barrier to escape. I personally think he got what he deserved for cutting his science budget.

The key breakthrough in describing the energy gap was not Bardeen's but Schrieffer's. Bob, who was only twenty-five in the winter of 1957, recounts that he was in New York attending a scientific meeting and got the idea on the subway.[10] This story is too wonderful to have been made up, for anyone who rode the New York subway in those years (it has since improved) knows that one's thoughts down there tended to be dark. The back of his brain must have been somewhere up in the sunlight chewing on the problem. What came into his mind was a mathematical description of the superconducting state so simple you could explain it in fifteen seconds. It was not a description of the real superconductor, of course, but a highly idealized abstraction of one containing the essentials—and, as it turns out, enough detail to account for key experimental findings. A modern version of Schrieffer's invention would be the computer game *Sim City*, a toy model of a real city you can play with and manipulate as you see fit, and which has enough in common with a real city that it teaches you some principles of how cities work. The theory of superconductivity has considerably more gravitas than *Sim City*, however, because it is falsifiable and wonderfully elegant. The confusion had seemed so intractable. When measured with a blunt instrument, such as a laser, the superconductor was indistinguishable from a sea of independent electrons, yet when measured with a subtle one, such as a pair of wires or a nearby magnet, it acted like the complete opposite, superfluid helium. It was like the transformation of absolute democracy at the scale of individuals, with its wild diversity of political opinions, into an absolute political party at the scale of nations, in which all traces of individuality had disappeared and been replaced by a single coherent message. But the problem turned out to have a simple technical solution that came to Bob on the train. He says he struggled all afternoon to articulate the idea and managed to write it down only later that evening—a revealing admission, for good theoretical physics is actually more like art than engineering and is similarly difficult to summon up on demand. The physical idea precedes the mathematics, and the act of writing it down as a simple equation is like capturing a song or a poem.

Students of superconductivity are often thrown by Schrieffer's equation because it is not the solution to any specific mathematical problem—other than contrived ones invented after the fact. It is conceptual, not technological, and is more an attempt to describe what occurs in nature as simply as possible than an attempt to master the world through mathematical deduction. The poor students must suddenly morph from logicians to contestants on the game show *Jeopardy* near the end of play when all the easy selections are gone, and the only option left is Hegelian Surprises for $500. Alex Trebek reads smoothly from his cue card, "The Bardeen–Cooper–Schrieffer theory of superconductivity," and exhorts the contestants to synthesize the corresponding question before the buzzer sounds.[11] There is, unfortunately, no way around this problem. Thinking through Schrieffer's idea initiates a sequence of events in which students discover, to their dismay, that real physics is nearly always inferential, that *no* collective organizational phenomenon—even such elementary things as crystallization and magnetism—has ever been deduced, and that the view to the contrary they learned when they were younger was simply a trick to get them to study. Superconductivity is not especially hard to grasp. It is merely the first case one encounters in which the ruse of mathematical deduction is so obvious that it can no longer be sustained. Schrieffer's idea was an immense intellectual triumph precisely for this reason. He had been indoctrinated just like the rest of us but somehow managed to overcome his training and get to the bottom of things. Indeed, the mistaken belief that superconductivity was a technological problem is why no one previously had managed to solve it!

The core of Schrieffer's idea was relaxing the particle number. This concept has a simple analogy in cities. Suppose you severed all the bridges and tunnels into Manhattan so that no one could go in or out. Life would go on more or less as usual with people moving about and repelling each other (I love New York) because the island is sufficiently large that one part of it functions as a reservoir for another. This contrasts sharply with, say, a crowded office party, where there can be a world of difference between having a door to the outside open or closed. If you now imagine Manhattan to be a piece of metal and the people in it to be electrons, Schrieffer's solution was simply to open the bridges and tunnels so as to let the number of electrons vary. In other words, since the number of electrons in any one area could vary without changing the properties of the whole, the number of electrons in *all* the areas could be allowed to vary as well, even though in reality this number was fixed. Allowing such variations is a standard mathematical trick used to simplify the description of conventional hot gases and liquids, but its use in superconductors was quite radical because the superconductor is ice cold. In hot, real-life Manhattan the number of people varies from moment to moment but is fixed at any particular time. In icy Schriefferesque Manhattan, in contrast, the number of people would be *undefined* and the quantum wave function of the city a lifeless, time-invariant mixture of states with different numbers of people. This is quite a concept. The simultaneous presence of classically incompatible things, in this case different numbers of electrons, is something Schrieffer's superconductor has in common with Schrödinger's cat.

The mathematical device of allowing the number of electrons in the sample to vary turns out to have important physical content, although Schrieffer did not realize it at the time. He was simply trying to generalize a technical idea that his colleague Leon Cooper had about instability of the electron sea. We now understand that he had accidentally stumbled upon a terse description of the violent quantum-mechanical sloshing of electrons from one part of the sample to the other characteristic of the superconducting state. It is possible to describe this effect without violating the particle number, but the resulting gain in rigor is vastly outweighed by the loss of clarity, and moreover misses the point. Superconductivity is an organizational phenomenon that, like crystallization, is *undefined* when the number of electrons is too small. The failure of Schrieffer's approximation in a small sample has the simple physical meaning that superconductivity cannot occur in such samples.

The number uncertainty required for Schrieffer's idea to work has a bizarre side effect that was initially hushed up but later understood to be crucial: the description of the superconducting state is not unique. There is an enormous number of equivalent solutions—about one quintillion in a one-centimeter cube of lead—each of which is as valid as any other.[12] This multiplicity is at first immensely troubling, because the microscopic rules of quantum mechanics require the state of a system to be unique. It was one of the main reasons the theory of superconductivity took so long to become accepted. But the effect is not that difficult to understand if you examine it in the proper light. There is only one history of the Roman Empire, but many of the fine details, such as who bought what decorative tile on what day for which villa atrium, could have been changed without affecting any important large-scale event. The number of plausible histories of the Roman Empire that make sense and get the essentials right is actually staggering. Histories of large systems are simply different from those of small ones because they are descriptions of collective phenomena and not of pedantic detail. The effect being emulated in the theory of superconductivity is like this. It is the tendency of the electrons to lock arms and move as one gigantic body, just as crystallized atoms do. It is actually no different from what happens in crystallization, except in being more difficult to cover up by skipping to the "obvious" nonquantum description at key moments. When the number of electrons is extremely large, it becomes difficult to distinguish the true ground state of the superconductor from the low-lying excited states associated with collective motion of the entire assemblage. The nonuniqueness of Schrieffer's description is thus a symptom of something extremely fundamental: the emergence of conventional meaning of the fluid body—the collective effect that transforms quantum mechanics into Newton's laws. It is interesting that many physicists continue to be confused about this matter, thus demonstrating that youth is not a prerequisite for getting intellectually mugged by nature.

Superconductors exhibit a number of exact behaviors that owe their exactness to Schrieffer's emergent multiplicity of ground states.

The most celebrated of these is the Meissner effect, the spontaneous levitation of a small piece of superconductor placed above the pole of a magnet. This levitation comes and goes reversibly as the sample is heated and cooled through its superconducting transition, and is therefore fun to demonstrate in class. Having seen hundreds of movie special effects, students nowadays are often desensitized to physical miracles, but this changes the moment they see Meissner levitation. The Josephson effect, actually two phenomena with the same name, is also quite breathtaking. One effect is the ability of a superconducting lead sandwich to conduct electricity without applied voltage. This is the physical basis of the ultrasensitive magnetism detectors with the endearing name *squid* (superconducting quantum interference devices) used in antisubmarine warfare, magnetic-resonance imaging, and magnetoencephalography. The other effect is the previously mentioned emission of radio waves when a voltage is applied to the sandwich. The proportionality constant between the timing of these waves and the applied voltage is constant from one experiment to the next to one part in a billion. Like the von Klitzing effect, the Josephson effect was predicted theoretically, but its extreme reproducibility was not. The Josephson constant is also a combination (although a different one) of the fundamental electric charge quantum e, Planck's constant h, and the speed of light c, and may thus may be combined with the von Klitzing constant and an independent measurement of the speed of light to yield e and h. Indeed, these two macroscopic effects are the practical present-day *definition* of these ostensibly microscopic quantities. The constancy of the Meissner and Josephson effects amounts to experimental proof that a principle of organization is at work in superconductors, one we now identify with Schrieffer's multiplicity and call superfluid symmetry-breaking.

The exactness of these effects brings up an epistemological matter that I can best illustrate with a story. Once when I was a kid I lost an honesty contest. I was swimming with one of my cousins in a large, spring-fed pool among the pines near my grandmother's cabin in the mountains above Porterville in California. It was fairly rugged country, and the road back to my grandmother's place was unbearably long and twisted, for it had to follow the contours of the canyon. On foot, however, one could take a nice shortcut trail down by the river. It was getting toward suppertime, and my cousin, who had been with me all day and was beginning to find me exasperating, initiated a metaphysical argument over whether the shortcut was actually short. He was a worldly wise fellow from the big city whose opinions I valued enormously, so I rose to the bait and vowed to prove him wrong, which he clearly was. He asserted confidently that I did not know what I was talking about, as usual, and that we should resolve the matter by racing back the two ways separately and seeing who arrived first. But the river trail was steep and full of rocks and tree roots, and thus difficult to negotiate with flip-flops at a run. So we negotiated a compromise in which both of us would *walk* home as fast as we could, and not cheat. We then set off. I dealt with all those inclines, tree roots, and willow boughs like a man, and got the stubbed toes and scratches to prove it, but when I finally arrived at the cabin, puffing up the trail, there was my cousin with a decimated corn cob and already halfway through his steak. He had won the contest, he said, and moreover, I was in big trouble for being late. I am ashamed to say how long it took me to figure out that he had simply run. Scientists are really very gullible people, and I am afraid I revealed my professional destiny by admitting defeat, faulting my own judgment, and going back over the facts again and again looking for my mistake. What a sap.

Scientific contests, like our footraces, are often won for the wrong reasons. The fight over the theory of superconductivity was one of the longest and bitterest in the history of science, primarily because the central issue was conceptual. The theory was eventually accepted on the basis of the "spectroscopic" detail it accounted for—the heat capacity (worked out by Bardeen at dinner), the heat transport coefficient, the energy gap, the relationship of this gap to the superconducting transition temperature, the variation of this temperature with isotopic mass, the modifications of the speed of sound that occurred at the transition, and so forth. The machinery of science is, sadly, not set up to deal with concepts but only with facts and technologies. Thus the Bardeen–Cooper–Schrieffer theory has been incorporated into the body of science not as a concept but as a computational technology. The cognoscenti understand the essential thing to be the exactness of the Meissner and Josephson effects, neither of which requires the rest of the theory to be true, but textbooks still tell the story through the spectroscopic detail the theory explains—and always will. Thus they say that superconductivity is an instability of the electron sea. They say that the attractive force between electrons causing this to happen is mediated by atomic motion. They say that the superconducting state has an energy gap related in a simple way to the transition temperature. And so forth and so forth.

In fact, none of these things is essential. It was just a historical accident that the first superconductors to be discovered fit spectroscopic details of the theory well and thus could be used to justify it. But the Meissner and Josephson effects are what actually justify the soul of the theory, not all those details. This fact was deeply understood from the beginning by a number of excellent Russian physicists, who feel to this day, with some justification, that they were unfairly scooped on the theory. Unhappily, life is unfair, notably and especially in the matter of concepts. Whenever my students get depressed about this state of affairs I just remind them of the words of Dr. Pangloss as he lay dying of syphilis.[13] When asked whether the Devil was at fault, he answered that the disease was inevitable in this Best of all Possible Worlds because it was brought to Europe by Columbus, who had also brought chocolate and cochineal.

The mental compromises required to define the theory of super-conductivity as a technology had the side effect of generating deep cultural confusion over the relative importance of things. Back in the 1970s two highly respected theoretical physicists (who will remain unnamed) recorded the contemporary prejudices of their discipline by writing a paper "proving" that superconductivity could never occur at temperatures higher than 30 degrees Kelvin (30 degrees above absolute zero). This was fully consistent with known properties of metals at the time and with the details of the theory of superconductivity fit to them. It was also important, since refrigerating things below 77 degrees Kelvin, the boiling point of liquid nitrogen, is especially expensive and thus an impediment to technology. Then, in a truly heartwarming turn of events, Georg Bednorz and Alex Müller discovered 30-degree-Kelvin superconductivity in a ceramic material that should not even have been metallic in the first place,[14] and shortly thereafter, Paul Chu discovered a similar material exhibiting 90-degree-Kelvin superconductivity. These sudden and disturbing developments caused a frenzy of creative backpedaling similar to the moment in a Road Runner cartoon when Wile E. Coyote discovers that his Acme rocket-powered sled has overshot the cliff. One heard all sorts of wild excuses, including suggestions that the phenomenon was not superconductivity at all but some fundamentally new kind of collective effect—conveniently absolving it of the need to conform to the Bardeen–Cooper–Schrieffer motif. But of course it wasn't. The experiments eventually became reproducible and clear, and the Josephson effect was made to occur between a high-temperature superconductor and a regular one, thanks to a clever surface preparation technique invented by Bob Dynes, now the president of the University of California. The mystery dissolved. What had failed was not the fundamental nature of the superconducting state, which was conventional, but the mythological electron sea on which it was ostensibly built. The materials in question simply did not have one.

The human side of high-temperature superconductivity is complicated, as is often the case with collapsing ideologies. There are still vicious fights similar to the medieval controversies over how many angels fit on the head of a pin, and valiant attempts to invent technical mathematics that would "explain" these superconductors the way the original theory explained conventional ones. But the sad truth is that the mathematics of the original Bardeen–Cooper–Schrieffer theory was not important in itself but only as a means of demonstrating the existence and nature of a new kind of order. Now that this order has been shown to exist and the new superconductors have been shown experimentally to exhibit it, there is no compelling reason to invent such a computational technology—other than perhaps for engineering purposes. On that score, Lev Landau, the famous Russian theorist responsible for first clearly codifying the properties of the electron sea, once said that you can calculate the properties of water, but it makes so much more sense just to measure them.

The reductionist response to high-temperature superconductivity reminds me of what the *New York Times* recently reported as the world's most popular joke.[15] Sherlock Holmes and Dr. Watson are on a camping trip:

Holmes: Watson, look up at those stars in the sky! What do you deduce?

Watson: Well, each of those pinpricks of light is a huge sun powered by the fires of hydrogen fusion. That fuzzy patch over there is the Andromeda galaxy. Powerful telescopes tell us that Andromeda is an island of billions and billions of stars. Even more powerful telescopes tell us that there are billions and billions of such galaxies stretching out to the edge of the universe. If even one in a million of those suns had planets, and even one in a million of these had an oxygen atmosphere, and even one in a million of these had life, and even one in a million of these had people and civilizations, then we would be certain of not being alone in the universe.

Holmes: No, Watson, you idiot! Someone stole our tent!

Reductionist ideology has another fascinating manifestation in superconductivity theory that I call quantum field theory idolatry. Quantum field theory, a body of mathematics that grew out of studies of elementary particles, is commonly taught after conventional quantum mechanics as a special language for working in that subject—and also as kind of superior way of thinking. It is actually not a new way of thinking but simply a restatement of quantum mechanics in the context of the special limitations and conditions appropriate to the vacuum of space. These conditions make the formalism elegant and fun to learn—at least for mathematical types such as myself—but they also make it easy to hide the essence of a thing through manipulation. Sleights of hand can make some physical behavior appear to be caused by field theory when it is actually being caused by the manipulation itself. Shortly after the Bardeen–Cooper–Schrieffer theory was introduced, the language of quantum fields was discovered to be particularly well suited for describing important properties of superconductors—notably the supercurrents themselves, the Meissner effect, the above-threshold conductivity, and the collective sloshing motions of the electrons called plasma oscillations—precisely because it allowed one to postulate one's way around messy, ultimately unimportant detail and get quickly to the meat of the matter. This eventually led to the practice of explaining all of superconductivity using field theory, and thus indirectly to the idea that quantum fields cause superconductivity. Even today one will find a great many people who secretly believe this. It is ridiculous—like believing that the weather is caused by the price of corn. In fact, quantum field theory works because the emergent universality of superconductivity makes it work, not the other way around. The microscopic equations of quantum mechanics encrypted in the field theory are different from those of the real material and therefore wrong. The only way one can start from wrong equations and get the right answer is if the property one is calculating is robustly insensitive to details, i.e., is emergent. Thus the lesson from superconductivity is actually not that quantum field theory is a superior computational technology but that quantum fields can themselves emerge.

The logical inconsistency of these two traditions reflects the depth of the crisis caused by the solution of the superconductivity problem—a confrontation between reductionist and emergent principles that continues today—and thus the monumental nature of the solution itself. It is said that Cooper found the mechanism, Schrieffer found the solution, and Bardeen realized why the solution was right. Of the three, the last was clearly the most important, which is why John Bardeen is so honored among physicists.

In modern times it is customary to view Bill Gates, the savvy businessman, as the ultimate technologist, but I think the real hero of the electronic age was John Bardeen. Bardeen always flew tourist class and did not like to think about his Nobel Prizes. A colleague once recounted how he was visiting the Bardeen home as a student when someone asked to see the transistor medal. His host could not remember where it was at first, and then rummaged around and eventually found it at the bottom of a sock drawer. Richard Feynman, the inventor of the most elegant aspects of quantum field theory, recounts how he was working on superfluidity and superconductivity when the Bardeen–Cooper–Schrieffer paper arrived in the mail as a preprint. He put it in a drawer unread and could not bring himself to read it for several months.

I once had an interaction with John Bardeen in which I came off as an arrogant young whippersnapper. I am not especially proud of this story, but I will tell it nonetheless because I know John would find its

Freudian overtones hilarious.[16] It was at a conference on many-particle quantum physics in northern Sweden in a hunting lodge far from anything but mountains and some lichen bogs. I had just come from another conference in Beijing and had picked up some awful intestinal ailment, so I was keeping everyone up when they wanted to sleep. My nocturnal comings and goings added to the problem of the light, which was a terrible nuisance because the sun was dipping into twilight only two hours before midnight and rising again four hours later.[17] There was a bit too much reindeer on the menu—roast reindeer, reindeer meatballs, pickled reindeer, and so forth—but then Sweden is often like this. At any rate, the key moment came at the end of the second day's activities in a slot marked "cocktail party" on the program. There was an extremely loud noise outside, and we all rushed out to discover parked on the lawn two industrial-strength helicopters piloted by top guns from the Swedish air force. These helicopters proceeded to ferry us in parties of six a few kilometers back into the mountains to the edge of a small lake carved into the granite by glaciers. There someone had erected a small tent and built a big campfire with some fortified mulled wine cooking on it called something like "wolf urine." So we stood there in the bright Arctic evening watching the breeze play with the fire in that desolate place, drinking mugs of this stuff, and opining how no conference before or after could possibly top this one. Meanwhile, I was growing more and more uncomfortable. Then it came time to helicopter back, and six of us piled in for the ride home. When we landed there was a brief moment when nobody did anything, so I started to push my way out. My colleague Gerry Mahan, who was sitting in the back with me, held me back violently, as though I was about to commit the greatest sin in the world. An old gentleman in the front slowly stretched his foot through the door and climbed out. It was John Bardeen.

NINE

The Nuclear Family

But this is an old and everlasting story: what happened in old times with the Stoics still happens today, as soon as ever a philosophy begins to believe in itself. It always creates the world in its own image; it cannot do otherwise.

F. Nietzsche

ONE OF THE STRANGER DEVELOPMENTS OF MODERN LIFE is the mythologization of nuclear weapons. It is a generational effect I have observed through talking with students from all sorts of backgrounds, and also from my own sons. They find the human side of warfare extremely difficult to understand and so think of these things as abstractions of power rather than agencies of killing. This was driven home to me when I took my older son to the atomic bomb museum in Hiroshima last summer. He found the horror of it literally incomprehensible, and outside by the famous skeletal atomic bomb dome he could see only the street musicians playing guitars down by the river's edge and the people his own age in wet suits zooming by on Kawasaki jet skis. The more the use of nuclear weapons recedes into the past, the more this technology turns unreal in the consciousness of civilization, like blasting spaceships and transfiguring robots on Saturday morning cartoon shows.

Nuclear weapons are, unfortunately, the most sensational engineering contribution of physics, something that catapulted the discipline to prominence in the 1950s and has colored it indelibly ever since. This coloring is inherently reductionist. The discovery of radioactivity and the subsequent study of nuclear reactions led to development of nuclear energy. This, in turn, led to the popular practice of conceptually subordinating everything to the laws of atomic nuclei—an effect caused at least partly by the stupendous amounts of money invested in physics after the war in tacit support of nuclear weapons.[1] It is natural for a one's worldview to be influenced by how one makes a living, and the tail is just as capable of wagging the dog in science as anywhere else.

Ironically, the physical principles involved in nuclear weapons are neither subtle nor sophisticated. At the Livermore Laboratory, where I used to work, there was continual talk about eliminating nuclear physics research on the grounds that it was irrelevant to nuclear weapons. Nuclear explosions are like fire. Once you've assembled the fuel, you need only ignite the reaction for it to run away and blow up. This is the truly scary thing about this technology and the reason governments the world over are so paranoid about the proliferation of fissile fuel. Once you have the parts, you can build a nuclear weapon fairly easily.

It was discovered back in the 1930s that radioactivity was a lot like chemistry except for the scales.[2] The nucleus of an atom is about a million times smaller than the atom itself and releases a million times more heat per reaction. The reactions themselves involve parts of the nucleus flying off, capture of nearby electrons by the nucleus, and consolidation of two small nuclei into a larger one, all of which have chemical analogues that occur in fire. These processes also obey the same laws of quantum mechanics that operate in chemistry. The one significant difference is that the forces among the various parts of the nucleus are not simple. In chemistry one has elementary electric force and nothing else, but in nuclei one also has forces that defy simple description and go by the inspired name of nuclear force.

Nuclear force is typically a student's first encounter with the idea that empty space is not really empty. Coming to grips with this fact—a physics rite of passage—is simultaneously thrilling and upsetting, like sneaking off to a dark place with your girlfriend and discovering that you have mistakenly gone to the bunkhouse. Even though there are other people snoring away in the bunkhouse, it is essentially just the two of you in there, although your behavior has been modified. Similarly, the inside of a nucleus is essentially protons and neutrons, but their behavior is modified by the medium through which they move, the ostensibly empty vacuum of space. In both cases the medium is passive only when one does a delicate experiment, such as one involving whispering and tiptoeing. In that case things conform to a theory in which there are nothing but the principal actors and interactions between them, however unusual and complicated. But in a violent experiment, the dynamical nature of the medium becomes visible, and all such theories fail.

Violence is commonplace in nuclear physics because the forces between protons and neutrons are so enormous. Attempts to perform a delicate experiment on a nucleus inevitably wind up like that Gary Larson cartoon, "The Pillsbury Doughboy meets Frank's Asphalt and Paving Service." The usual practice is not to try, but instead to whack one nucleus with another one at great speed and see what flies out. Ironically, one of the few delicate effects in nuclear physics turns out to be thermal uranium fission. An amazing accident of nature enables a neutron traveling no faster than a common air molecule to initiate the reaction, thus amplifying the energy of the neutron by a factor of one hundred million. This special property of uranium is what makes water-mediated nuclear reactors feasible.

Like many other people, I have some personal experience with spaces that are supposed to be empty but are not. In the early 1970s when I was a soldier, I decided to go camping in Switzerland over the weekend with another fellow from my unit. For convenience we decided to take the train, but we managed to miss a connection in Stuttgart and arrived in Zurich so late at night that there were no more local expresses out to the trailheads. There was no way to book a room at that hour, so, in a fit of bad judgment I regret to this day, we went across the street to a park, unrolled our purple sleeping bags, and slept on a bench. I mean this figuratively, of course, for we did not actually sleep very much. We were "not alone." It continued to be very busy indeed in that dark park all night long, and I have never been so happy to see the sun as I was the next morning. Anyone doubting that empty space is a fiction should spend a night in a park.

The second, more direct, encounter with the problem of space is the strange ability of nuclei to create particles. One of the common forms of radioactivity, beta decay, involves emission of a high-speed electron and an accompanying antineutrino, a ghostly object capable of passing entirely through the center of the earth without hitting anything. One explains this effect by saying that a neutron, one of the constituents of a nucleus, "converts" to a proton, the other constituent of a nucleus, plus an electron and an antineutrino, which then escape. This explanation is consistent with the properties of a free-standing neutron liberated from the nucleus, which will "convert" in just this way in about a minute. In so doing it will also "convert" a person to a cancer patient if he is not careful—which is why it is usually better to send a graduate student down to fix a neutron spectrometer rather than going oneself. Thus one might be induced to describe the neutron as a bound state of a proton, electron, and antineutrino that comes apart of its own accord the way an unstable atom or molecule might in a chemical reaction. Unfortunately, there turns out to be a second kind of beta decay in which a proton in the nucleus "converts" backward into a neutron, throwing off an antielectron and a neutrino in the process. Thus it is neither correct nor very helpful to think of the neutron as comprising a proton plus other things. What has subverted the chemical analogy is the existence of antiparticles—versions of conventional particles but with opposite properties—and the ability of the nucleus to pop particle–antiparticle pairs out of the vacuum any time it likes, provided it can ante up the requisite energy. Which kind of beta decay one gets thus turns out to depend on the energy budget of the nuclear force, which likes to have a slight excess of neutrons over protons in the nucleus, but only just.

Antimatter is one of those bizarre facts of nature that is too crazy to have been made up by science fiction writers. It is an exact copy of ordinary matter except with all the electric charges reversed, so that it can react with ordinary matter in a catastrophic explosion that annihilates everything and leaves behind an angry swarm of gamma rays—the nuclear physicist's term for short-wavelength light. This explosion is the effect that powers the starship *Enterprise* in *Star Trek*. I always felt there was insufficient respect paid those gammas on *Star Trek*, and that the poor engineers down below should have been issued lead underwear for realism. Maybe this explains where all the aliens came from. Unlike *Star Trek*, however, antimatter is real. It is created every day by radioactive decay and large accelerator laboratories all over the world.

The existence and properties of antimatter are profoundly important clues to the nature of the universe. Back in the 1920s it was discovered that writing quantum equations of motion for an isolated particle that could correctly describe its measured behavior at both low and high speeds was fundamentally impossible. The simplest solution—and the one that turned out to be experimentally correct—was to describe space as a system of many particles similar to an ordinary rock. This is not a precisely correct statement, since Paul Dirac formulated the relativistic theory of the electron before electrons and holes in crystalline solids were understood, but in hindsight it is clear that they are exactly the same idea. Thus in elemental silicon, where there are many electrons locked up in the chemical bonds, it is possible to pull an electron out of a chemical bond to make a hole. This hole is then mobile, and acts in every way like an extra electron added to the silicon, except that its electric charge is backward. This is the antimatter effect. Unfortunately, the hole idea makes no sense in the absence of something physically analogous to a solid's bond length, since this length fixes the density of electrons one is ripping out. Without it, the background electron density would have to be infinity. However, such a length conflicts fundamentally with the principle of relativity, which forbids space from having any preferred scales. No solution to this dilemma has ever been found. Instead, physicists have developed clever semantic techniques for papering it over. Thus instead of holes one speaks of antiparticles. Instead of a bond length one speaks of an abstraction called the ultraviolet cutoff, a tiny length scale introduced into the problem to regulate it—which is to say, to cause it to make sense. Below this scale one simply aborts one's calculations, as though the equations were becoming invalid at this scale anyway because it is, well, the bond length. One carries the ultraviolet cutoff through all calculations and then argues at the end that it is too small to be measured and therefore does not exist.

The ultraviolet cutoff problem reminds me of the scene in Mel Brooks's *Young Frankenstein* where Dr. Frankenstein asks his hunchbacked servant, Igor, how he lives with his hump, and Igor answers, "What hump?" Much of quantum electrodynamics, the mathematical description of how light communicates with the ocean of electrons ostensibly pervading the universe, boils down to demonstrating the unmeasurableness of the ultraviolet cutoff. This communication, which is large, has the fascinating implication that real light involves motion of something occupying the vacuum of space, namely all those electrons (and other things as well), although the extent of this motion depends sensitively on the value of the ultraviolet cutoff, which is not known. There are endless arguments about what kinds of regularization are best, whether the cutoff is real or fictitious, whether relativity should be sacrificed, and who is too myopic to see the truth. It is just dreadful. The potential of overcoming the ultraviolet problem is also the deeper reason for the allure of string theory, a microscopic model for the vacuum that has failed to account for any measured thing.

The source of this insanity is easy to see if one simply steps back from the problem and examines it as a whole. The properties of empty space relevant to our lives show all the signs of being emergent phenomena characteristic of a phase of matter. They are simple, exact, model-insensitive, and universal. This is what insensitivity to the ultraviolet cutoff means physically.

The similarities between the vacuum of space and low-temperature phases of matter are legendary in physics. Not only are phases static, uniform quantum states, but their most subtle internal motions are physically indistinguishable from elementary particles *very generally.*[3] This is one of the most astonishing facts in science, and something students always find upsetting and difficult to believe. But they eventually become convinced after looking at enough experiments, for the evidence is plentiful and consistent. In fact, the more one studies the mathematical descriptions of cold phases, the more accustomed one gets to using the parallel terminologies of matter and space interchangeably. Thus instead of a phase of matter we speak of a vacuum. Instead of particles we speak of excitations. Instead of collective motions we speak of quasiparticles. The prefix "quasi" turns out to be a vestige of the historical battles over the physical meaning of these objects and conveys no meaning. In private conversations one drops the pretense and refers to the objects as particles.

Zero-temperature phases are not very charismatic things, unfortunately, at least on the surface, so people's obsession with them is an easy target of techno-humor. When I was a student in the mid 1970s, for example, I heard a satire on them modeled after that awful article the *National Lampoon* ran on Dan Blocker, the actor who played Hoss on the television program *Bonanza*. Dan had just died of a pulmonary embolism, and somebody on the editorial staff decided it would be funny to "interview" him and ask all sorts of questions about the show, events of the day, the latest movies, and so forth, to which he would continually respond with silence. In the spoof, Dan was reincarnated as a tank of helium–3, and the same interviewer asked him all about his new life, how it felt to go with the flow, whether he ever got excited, how he was taking the pressure, and so forth. This occurred at MIT.

From a more thoughtful perspective, however, the obsession is not so funny. Individuals have worked on these systems with immense dedication, in some cases jeopardizing their personal financial security, for zero-temperature phases (other than semiconductors and ordinary metals) have generated little economic value and are despised by funding agents and investors as a result. But a felicitous consequence is that this body of work is unusually reliable, for it has been a labor of love performed with great care and openness. It is the source of our confidence that the analogy between antiparticles and holes in ordinary crystalline insulators is exact, robust, and universal. It is how we know that the analogy extends to superconducting metals and superfluid helium–3, a perfectly homogeneous substance lacking crystalline order.[4] It is how we know that the superfluid liquid and gas both exist,[5] and that matter inside an atomic nucleus is a fluid. This last part is the conceptual basis of our understanding of neutron stars[6] and the crust of quantum liquid crystal phase that may form at their surfaces.[7]

After electrons and holes, the simplest example of emergence of particles in rocks is sound quantization. This astonishing phenomenon is the closest thing to real magic I know. Sound is familiar to everyone as the vibration of elastic matter, typically air but also solid walls, as you know if you've attempted to sleep with a loud party going on next door. Of the two, sound in solids is the more interesting from a quantum perspective because it continues to exist and make sense even at ultralow temperatures. Measurements at such temperatures reveal that it is particulate. Suppose, for example, a sound transducer is attached to a solid and turned on, thus beaming sound into the solid, and then reduced in intensity to make the amount of sound small. A sound receiver on the other side of the solid detects not a faint tone but sharp pulses of energy arriving at random times. This quantized transmission of pulses evolves into the more familiar transmission of tone when the intensity is increased— an everyday example of the emergence of Newtonian reality out of quantum mechanics. But at low intensities this emergence does not occur, and the conclusion becomes inescapable that particles of sound exist, even though they do not exist when the solid is disassembled into atoms. The particles emerge, just as the solid itself does.

Sound quantization is a particularly instructive example of particle emergence because it can be worked out exactly, in all its detail, starting from the underlying laws of quantum mechanics obeyed by atoms—provided the atoms are first postulated to have perfectly crystallized. This is what we mean by quantized sound being a universal feature of crystallinity. This phenomenon is the prototypical example of Goldstone's theorem, the statement that particles necessarily emerge in any matter exhibiting spontaneous broken symmetry. The analysis also reveals that the particles of sound acquire more and more integrity as the corresponding tone is lowered in pitch, and become exact in the limit of low tone. Very high-pitched sound quanta propagating through a solid can decay probabilistically into two or more quanta of sound with a lower pitch, this decay being aptly analogous to that of a radioactive nucleus or an elementary particle such as a pion. Their decay turns out to be the same thing as elastic nonlinearity—the failure of distortion of the solid to be proportional to the stress on it when the stress is large, such as occurs just before fracture. But since these nonlinearities matter less and less as the sound wavelength increases, the time scale for the decay increases as the tone is lowered and eventually becomes infinite. Sound quantization is a beautiful case of magic in physics revealed by thoughtful analysis to be not magic at all but a failure of intuition.

The quantum properties of sound are identical to those of light. This fact is important, for it is not at all obvious, given that sound is a collective motion of elastic matter while light ostensibly is not. The analogy is revealed most simply and directly by heat capacity. The ability of crystalline insulators to store heat drops universally in cryogenic environments as the cube of the temperature. This effect is a consequence of quantum mechanics, for it is easy to show that the heat capacity would have to be constant and large (as it is at room temperature) if all the atoms obeyed Newton's laws. The heat capacity of empty space follows the rule precisely. Space is not empty when it is heated, of course, but filled with light, the color and intensity of which depend on the temperature. This effect is familiar from the red glow emitted from hot embers and the white light blazing from a light bulb filament or the surface of the sun. A warm crystal is likewise filled with sound. In either case the specific temperature dependence of the heat capacity is accounted for quantitatively by the Planck law, a simple formula derived from the assumption that light or sound can be created or annihilated only in discrete amounts.[8] In fact, the formula for the heat capacity of a crystalline solid is simply that of empty space, with the speed of sound substituted for the speed of light. The emergent quantum of sound, known as a *phonon*, is aptly analogous to the quantum of light, the *photon*. The physical equivalence of these two kinds of particle has been confirmed by a large number of experiments, some quite beautiful and clever.[9]

The analogy between phonons and photons raises the obvious question of whether light itself might be emergent. Here one must be careful to separate the legitimate issue of the vacuum of space as a phase from the bogus one of whether it is a phase we know. The common argument that the vacuum is not a phase because it is not a solid (which it is not) is like saying a dying person does not have a disease because it is not smallpox. The phases of matter have not all been discovered, and they certainly cannot be deduced from first principles. This is true even in the world of everyday chemistry, and all the more so in the much larger world of possible microscopic underpinnings of the universe. To think productively about the matter we must focus on what we know and not over-theorize. The similarity between sound and light requires explanation, for there is no obvious reason for their quantum mechanics to be the same. In the case of sound, quantization may be deduced from the underlying laws of quantum mechanics obeyed by the atoms. In the case of light it must be postulated. This logical loose end is enormously embarrassing, and is something we physicists prefer to disguise in formal language. Thus we say that light and sound obey the Planck law by virtue of canonical quantization and the bosonic nature of the underlying degrees of freedom. But this is no explanation at all, for the reasoning is circular. Stripped of its complexity, "canonical quantization" simply boils down to requiring light to have properties modeled after those of sound.

Light has a vexing aspect, the gauge effect, that has no analogue in sound and is often used to argue that light cannot be emergent. This argument is false, since there are plenty of ways one could imagine that light might emerge, but the effect is nonetheless a serious conceptual matter that points to an important physical distinction between light and sound. Its simplest manifestation is in heat capacity. When a sound wave passes by, a given atom is displaced a bit from its static position in the lattice. There are three distinct ways it can do this—left-right, up-down, and forward-backward—each of which contributes separately to the heat capacity, effectively multiplying the final answer by three. But the corresponding multiplication factor for light is only two, even though light is also the displacement of something. On one of the three axes the stuff of the universe, whatever it is, simply cannot vibrate—at least on time scales relevant to experimentally accessible temperatures—or store heat. The underlying microscopic reason is not known and is treated in modern physics as a postulate.

Simply defining away the gauge effect, however, has a number of unsavory aspects that look suspiciously like loose ends of a reductionist interpretation of an emergent phenomenon—rather like those bits of evidence you find behind the popcorn in the pantry revealing that you have mice. It turns out, for example, that making an entire mode of vibration disappear is difficult, especially when it involves motion of objects—the electrons in the vacuum—that vibrate perfectly well by themselves in all three directions. The trick is to postulate the quantum-mechanical wave function of the light to be entangled in a certain way with all the electrically charged matter in the universe—including that buried in the vacuum of space itself—before there has been any physical contact between them. Once established, this entanglement then persists throughout eternity and forbids certain things from happening. There is also a fundamental incompatibility of the gauge effect with the principle of relativity, which one must sweep under the rug by manipulating the ultraviolet cutoff. Finally, there is the problem that the "unphysical" motions of the waves encrypted in the entanglement begin life as physical ones in the mathematical description and become unphysical only at the end of the calculation by virtue of the fact that they cannot be measured.

A serious candidate for an emergent cause of the gauge effect is superconductivity. The well-known congruence between superconductivity and the gauge effect is implicated in the exactness of the Meissner effect, and is the reason the onset of superconductivity is often called gauge symmetry-breaking. It is also the reason superfluidity is a central component of most models in which a gauge principle might emerge. All such models constructed to date have been contrived and unsatisfying, not so much because they are wrong but because they are unfalsifiable. At the experimental scales accessible to us these models cannot be distinguished from one another, or from ones in which the gauge principle is simply postulated. The accepted practice of declaring unmeasurable things to be nonexistent—even when the problem lies in one's own experimental shortcomings—then makes the issue moot.

A much less controversial case of emergence of vacuum properties is the special relationship seen between the forces of electricity and nuclear decay on the one hand, and the masses of two special elementary particles, called W and Z bosons, on the other.[10] The physical idea behind this relationship is that a superconducting fluid—more precisely, a multicomponent abstraction of such a fluid—pervades the universe and modifies the electric force to create the weak nuclear force, somewhat as a laboratory superconductor modifies electric forces. This fluid also has sloshing motions, which, like sound in a solid, are quantized and thus show up experimentally as particles.[11] The corresponding sloshing motion of the superconductor, called a *plasmon*, is seen routinely in electron microscope experiments.[12] Not only are the W and Z bosons observed to exist, but their slight mass difference is exactly the value required by the observed differences between the nuclear and electric force strengths. Whether such a fluid really exists is still somewhat controversial because the Higgs particle, a more sophisticated sloshing motion of the fluid, has not yet been observed. The reason is almost certainly the technical limitations of existing accelerators, and most physicists expect the Higgs particle to be found soon.

Many other aspects of the vacuum look suspiciously emergent. There is, for example, the great simplicity of its quantum field-theoretic description, which is unusual because such descriptions in ordinary matter tend to be complicated except when they emerge—as they do in a superconductor or superfluid. There is also the hierarchy of scales, the tendency of phenomena at progressively longer lengths and times to be sequentially subordinate. The vacuum, when cooled from very high temperatures, is thought to undergo stepwise events called unification transitions, in which the known forces of nature split off sequentially from their fundamental parent. Similarly, holmium metal, a rare earth element, when cooled from very high temperature, first condenses into a liquid at 2993 degrees Kelvin, then solidifies at 1743 degrees, then develops a special kind of spiraling magnetism at 130 degrees, then tilts the spiral to make a weak ferromagnet at 20 degrees.[13] Between 130 degrees and 20 degrees the pitch of the spiral varies continuously, as though it were a rubber screw being stretched along its axis. With each of these transitions the "forces" between electrons in the metal transmitted by various elastic distortions of the ordered states split off from the fundamental parent in beautiful and apt analogy with what occurs in the vacuum. The temperatures required to see the unification transitions of the vacuum cannot be achieved in the laboratory, or even in the centers of the largest stars, so the evidence for unification is indirect, but it nicely parallels what one would find in the tilted spiral magnet if one's experiments could reach only long length and time scales. One of the strongest of these bits of evidence is renormalizability, an effect that causes the accessible measurements to be simple and redundant (one measurement predicts another) but that at the same time is incapable of revealing anything about the forces at the top of the hierarchical tree. There are many other such examples.

As nuclear energy began to be exploited in the 1950s and large accelerators were built to explore the workings of the nuclear force, it slowly became apparent that things were becoming more complicated at subnuclear scales rather than less. The gauge principle, relativity, and the general properties of antimatter continued to hold, but as the number of elementary particles began to proliferate, so did the rules for their interactions. None of these discoveries turned out to be particularly helpful in understanding atomic nuclei, much less atoms, nor is it possible today to compute the masses of the proton and neutron accurately from the standard model of elementary particles. The equations are just too complicated. Such complexity is, of course, sadly familiar in ordinary matter, for it is exactly what occurs in, say, a piece of silicon if you make the mistake of measuring it too violently. You zoom right past the subtle, universal simplicity of electrons and holes and begin measuring all sorts of interesting but ultimately irrelevant details associated with the chemistry. Another thing sadly familiar is the proliferation of particle masses and couplings in the vacuum that have precise values but do not seem to be related to each other in any simple way. It is exactly like the shelf after shelf of mind-numbing substance properties in reference sections of chemistry libraries. These data flow logically from a few basic things, as far as we know, but are just easier to measure and tabulate than they are to calculate.

Despite all this evidence that the reductionist paradigm in physics is in trouble, subnuclear experiments are still generally described in reductionist terms. This is especially curious considering that much of the thinking built into the standard model reflects the idea that the vacuum is a phase and that the laws of physics are reasonably simple and straightforward at the nuclear scale—but not beyond—because they are universal properties of that phase. Nonetheless, instead of low-energy universality, physicists speak of effective field theory. Instead of phases, we speak of symmetry-breaking. Instead of phase transitions, the unification of forces. The situation reminds me of a hospital where no one ever dies but instead experiences "negative patient care outcome" or "failure to achieve wellness potential."[14] In either case the confusion is ideological. The death of a patient is an unthinkable failure of the hospital's mission to preserve life. The subordination of understanding to principles of phase organization is a similarly unthinkable failure of one's mission to master the universe with mathematics. In situations that matter, mythologies are immensely powerful things, and sometimes we humans go to enormous lengths to see the world as we think it should be, even when the evidence says we are mistaken.

There is, of course, a more worldly way to put it. My colleague George Chapline likes to quote what he calls the First Theorem of Science, which he attributes to me, but which I remember distinctly as coming from him: It is impossible to convince a person of any true thing that will cost him money. We should probably rename it the First Theorem and drop the Science part.

A corollary of this theorem is that truth is sometimes relative. When I was a high school student I participated in a highly revealing political simulation. Our teachers divided us up into teams, each representing the government of a different fictitious country. Every team was given a set of instructions that included a world history, a mission statement, some military capabilities, and so forth, and assigned a table. The play consisted in passing messages back and forth in Diplomatic Pouches (little slips of paper) and giving speeches at the World Forum (an aluminum table in the front of the room). Direct personal contact was not allowed. My country, one of the smaller ones, was distinguished from the others by having rich uranium deposits. The larger countries had none. My task as president was to maximize the value of these deposits for my people by selling ore to all buyers, maintaining the balance of power so that neither of the larger countries could extort low prices. So we played for about two hours, delivering vapid speeches and trading messages that did not seem to make any sense, at which point one of the larger countries suddenly invaded me in the name of world peace and security. The war did not last very long, for my country was small, and I was deposed. It was a stunning betrayal—these were my *friends*. I do not know how the game came out because I was no longer in power and was thus able to go out for a sandwich. But after the game concluded, the organizers revealed the secret: each country had received a different world history. While I had been trying to maximize profits my opponents had been obsessing on those ore deposits as a matter of national security. Each of the bigger countries believed that the other was plotting to deny them access, and that my government was secretly its client. No wonder they invaded.

In science, as in anything else, the best antidote to the disease of mythology is a healthy dose of experimental reality. I have a colleague, Chung-Wook Kim, who was present in Hiroshima when the bomb was dropped. Kim is ethnically Korean and now heads the Korean Institute for Advanced Study in Seoul. His father was very pro-Japanese in the 1940s and was working in Hiroshima as an expatriate businessman. Kim was in fifth grade. All grades from fourth up had taken refuge in a temple about nine kilometers from ground zero, where they had classes in the morning, and in the afternoon collected edible weeds and had military training with bamboo sticks using Roosevelt and Churchill as targets. At about 8:15 in the morning he remembers an unearthly brilliant white light streaming through an upper window and illuminating the entire room, followed a bit later by a thunderous noise. The teacher tried to calm them down, but they were all excited by the sight of the colorful huge mushroom, which was pointed out by one of his classmates. The teacher then told them that Japan had invented a new antiaircraft weapon, which was believable because they had air raids almost every day. But late in the afternoon they heard on the radio that the US had dropped a super bomb, and they all broke down. Kim says he can never forget what he saw in Hiroshima, will not describe most of it, and has not been back since. As one got closer to the center of town the buildings were first more and more inclined away from the blast center, then flattened. The authorities were pouring gasoline on piles of bodies and burning them. One of his aunts had died instantly in a building collapse near ground zero. A cousin in the vicinity survived the blast, but within a month lost all his hair, went crazy, and died—the classic symptoms of radiation sickness. Another cousin had been biking across a small bridge when the bomb went off and was blown off unconscious into the shallow water below. When he awoke he discovered sun (radiation) burn on half of his body. He died a few years later in Korea of mysterious causes. This story has an especially poignant ending, for Kim later become a respected neutrino physicist and wrote an excellent textbook on the subject that is still widely used.[15]

The point of recounting this story is not to wallow in wartime horror but to gently remind the young, notably my own sons, that self-deception has consequences. Most of the time the effect is not so dire as warfare, but simply a degradation of the quality of life. These degradations include such happy institutions as road rage, divorce court, and excessively long faculty meetings.

The more important matter is that ideologies preclude discovery. All of us see the world as we wish it were rather than as it actually is, for it is in our nature, but we need to keep in mind that this is a design flaw of the human mind and resist it if we can. Seeing through ideologies and debunking them is what real science is all about. Mental life in general, actually.

The Fabric of Space-Time

The mathematics is not there until we put it there.

Sir Arthur Eddington

EINSTEIN'S THEORY OF RELATIVITY, ONE OF OUR MOST enduring cultural icons, is something everyone has heard of but few people understand.[1] Its inventor's image is recognized the world over as a symbol of cosmically transcendent intelligence and wisdom. In the popular imagination, relativity is a kind of deeper reality that only those blessed with extraordinary mental gifts can comprehend.

These otherworldly atmospherics are both excessive and inaccurate. The original version of relativity, the so-called special theory, is actually a law and a rather simple one at that, being not an equation of motion at all but a *property* of that equation, a symmetry. The most mature form of relativity is a speculative post-Newtonian theory of gravity motivated by this law.[2] Einstein, who discovered early in his career that the public was more interested in the mystical aspects of relativity than the physical ones, encouraged the growth of his image as a seer even though he was not a seer at all but a professional with a razor-sharp mind. However, Einstein's writing is characteristically well-reasoned, direct, and open. He was capable of being wrong, just like the rest of us, but he rarely hid his mistakes in abstruse mathematics. Most physicists aspire to be as clear as Einstein, but few of them succeed.

Symmetry is an important, if often abused, idea in physics.[3] An example of symmetry is roundness. Billiard balls are round, and this allows one to make some predictions about them without knowing exactly what they are made of, for example, that they will roll in straight-line paths across the table when struck with a cue. But roundness does not *cause* them to move. The underlying laws of motion do that. Roundness is just a special property that sets billiard balls apart from arbitrary rigid bodies and is revealed by the unusual simplicity and regularity of their motion. Symmetry is especially helpful in situations where one does not know the underlying equations of motion and is trying to piece them together from incomplete experimental facts. If, for example, you knew that all billiard balls were round and were trying to guess their equations of motion, you could eliminate certain guesses on the grounds that round things could not possibly do this. Situations of this kind are the rule rather than the exception in subnuclear physics. For this reason there is a tradition in physics of ascribing to symmetries an overriding importance even though they are actually a consequence, or property, of the equations of motion.

The symmetry of relativity involves motion.[4] Einstein and other early twentieth-century figures came upon this symmetry through thinking about electricity and magnetism, whose equations had just been worked out by James Maxwell and were rapidly leading to the invention of radio. Rotational symmetry requires the behavior of billiard balls on a round table to appear qualitatively the same regardless of where one stands on the perimeter. Relativistic symmetry requires their behavior to appear the same regardless of how one is *moving*. This idea is captured brilliantly by the famous Einsteinian thought experiment of a passenger on a train watching another train pass by. Einstein proposed that in the ideal limit—two trains passing each other in the vacuum of space—no measurement could determine which train was stationary and which was moving. That being the case, the equations of electricity and magnetism would have to appear the same on the two trains, and thus the speed of light must also be the same. One then encounters a logical contradiction unless some common ideas about simultaneity and measurement on the two trains are wrong. All of these musings and their fascinating logical implications, including the weight gain acquired by objects moving at high speeds and the equivalence of mass and energy, are now routinely verified in laboratories all over the world, and have passed into history as self-evident truth.

The story of Einstein's triumph is so romantic it is easy to forget that relativity was a discovery and not an invention. It was subtly implicit in certain early experimental observations about electricity, and it took bold thinking to synthesize these observations into a coherent whole. But no such boldness would be required today. An unsuspecting experimentalist armed with a modern accelerator would stumble upon the effects of relativity the first day and would probably figure the whole thing out empirically in a month. Relativity is actually not shocking at all. The ostensibly self-evident worldview it supplanted was simply based on incomplete and inaccurate observations. Had all the facts been known, there would have been no controversy and thus nothing for Einstein to prove. The popular view of relativity as a creation of the human mind is wonderfully ennobling but in the end incorrect. Relativity was discovered. Einstein's beautiful arguments notwithstanding, we believe in relativity today not because it ought to be true, but because it is measured to be true.

Einstein's theory of gravity, in contrast, was an invention, something *not* on the verge of being discovered accidentally in the laboratory. It is still controversial and largely beyond the reach of experiment.[5] Its most important prediction is that space itself is dynamic. The equations Einstein proposed to describe gravity are similar to those of an elastic medium, such as a sheet of rubber. Conventional gravitational effects result when this medium is distorted statically by a large mass, such as a star. When the source oscillates rapidly, however, such as when two stars revolve around each other in tight orbit, there is a new effect: outwardly propagating ripples of gravity. Conventional gravity is thus like the dimples under the feet of a water skimmer, and gravitational radiation is like the disturbances generated by the skimmer when it scampers away. There is much indirect evidence that the prediction of gravitational radiation is correct, the strongest being the steadily diminishing orbital period of the famous binary pulsar discovered by Joseph Taylor and Russell Hulse in 1975.[6] There is as yet no direct evidence. Detecting gravitational radiation directly is one of the key goals of modern experimental physics,[7] but most physicists are already persuaded by other evidence that Einstein's theory of gravity is probably correct.

It is ironic that Einstein's most creative work, the general theory of relativity, should boil down to conceptualizing space as a medium when his original premise was that no such medium existed. The idea that space might be a kind of material substance is actually very ancient, going back to Greek Stoics and termed by them *ether*. Ether was firmly in Maxwell's mind when he invented the description of electromagnetism we use today. He imagined electric and magnetic fields to be displacements and flows of ether, and borrowed mathematics from the theory of fluids to describe them. Einstein, in contrast, utterly rejected the idea of ether and inferred from its nonexistence that the equations of electromagnetism had to be relative. But this same thought process led in the end to the very ether he had first rejected, albeit one with some special properties that ordinary elastic matter does not have.

The word "ether" has extremely negative connotations in theoretical physics because of its past association with opposition to relativity. This is unfortunate because, stripped of these connotations, it rather nicely captures the way most physicists actually think about the vacuum. In the early days of relativity the conviction that light must be waves *of* something ran so strong that Einstein was widely dismissed.[8] Even when Michelson and Morley demonstrated that the earth's orbital motion through the ether could not be detected, opponents argued that the earth must be dragging an envelope of ether along with it because relativity was lunacy and could not possibly be right. The virulence of this opposition eventually had the scandalous consequence of denying relativity a Nobel Prize. (Einstein got one anyway, but for other work.) Relativity actually says nothing about the existence or nonexistence of matter pervading the universe, only that any such matter must have relativistic symmetry.

It turns out that such matter exists. About the time relativity was becoming accepted, studies of radioactivity began showing that the empty vacuum of space had spectroscopic structure similar to that of ordinary quantum solids and fluids. Subsequent studies with large particle accelerators have now led us to understand that space is more like a piece of window glass than ideal Newtonian emptiness. It is filled with "stuff" that is normally transparent but can be made visible by hitting it sufficiently hard to knock out a part. The modern concept of the vacuum of space, confirmed every day by experiment, is a relativistic ether. But we do not call it this because it is taboo.

How Einstein came to conclude that space was a medium is a fascinating story. His starting point was the principle of equivalence, the observation that all objects fall under the pull of gravity at the same rate regardless of their mass. This is the effect that causes astronauts in near earth orbit to experience weightlessness. The pull of gravity is not significantly smaller in low orbit than on earth, but the effect of this gravity is simply to make them and their spacecraft fall together around the earth. Einstein inferred from this effect (more precisely from versions of it he imagined in 1905 when there were no astronauts) that the force of gravity was inherently fictitious, since it could always be turned off by allowing the observer and his immediate surroundings to fall freely. The important effect of a nearby massive body such as the earth was not to create gravitational forces but to make free-fall paths converge. Astronauts falling straight down onto the earth (an unfortunate experiment) might at first think they were in deep space, but after a while would notice that objects traveling with them were slowly getting closer. This is because all the nearby free-fall paths are directed toward the center of the earth and eventually meet there. Einstein was struck by the similarity between this effect and the convergence of lines of longitude at the north and south poles. In that case, the tendency of some straight-line paths to converge is a consequence of the curvature of the earth—a medium made out of conventional matter. Then, in a flash of insight that leaves us breathless even today, he guessed that free-fall paths actually *are* lines of longitude on a higher-dimensional surface, and that gravity occurs because large masses stretch this surface and cause it to curve. He then made a second, masterful guess about the specific relation between mass and curvature known to us today as the Einstein field equations. These respect relativity and thus contain the same paradoxes of simultaneity found in the original version of relativity. For this reason they are more accurately described as a relation between stress-energy and the curvature of four-dimensional space-time. Their prediction that space can ripple in addition to stretching is a consequence of its obeying relativity, a symmetry of motion. It is consistent with our physical intuition, however, since it is basically the same thing as a propagating seismic wave on the surface of the earth generated by an earthquake.

The clash between the philosophy of general relativity and what the theory actually says has never been reconciled by physicists and sometimes gives the subject a Kafkaesque flavor. On the one hand, we have the view, founded in the success of relativity, that space is something fundamentally different from the matter moving in it and thus not understandable through analogy with ordinary things. On the other, we have the obvious similarities between Einsteinian gravity and the dynamic warping of real surfaces, leading us to describe space-time as a *fabric*. Bright young students inevitably pick up on this and ask the professor what moves when gravitational radiation propagates. They receive the answer that space-time itself does, which stops them cold. It is like learning that the surface of the sea undulates because it is an undulating surface.[9] Wise students do not ask this question a second time.

Their curiosity is, however, neither naive nor inappropriate. The closet of general relativity contains a horrible skeleton known as the cosmological constant. This is a correction to the Einstein field equations compatible with relativity and having the physical meaning of a uniform mass density of relativistic ether. Einstein originally set this constant to zero on the grounds that no such effect seemed to exist. The vacuum, as far as anyone knew, was really empty. He then gave it a small nonzero value in response to cosmological observations that seemed to indicate the opposite, and then later removed it again as the observations improved. A nonzero value is again in fashion due to the development of a new technique for measuring astrophysical distances using supernovae.[10] However, none of this adjustment addresses the deeper problem. Given what we know about radioactivity and cosmic radiation, there is no reason anyone can think of why the cosmological constant should not be stupendously large—many orders of magnitude larger than the density of ordinary matter. The fact that it is so small tells us that gravity and the relativistic matter pervading the universe are fundamentally related in some mysterious way that is not yet understood, since the alternative would require a stupendous miracle.

The view of space-time as a nonsubstance with substance-like properties is neither logical nor consistent with the facts. It is instead an ideology that grew out of old battles over the validity of relativity. At its core is the belief that the symmetry of relativity is different from all other symmetries in being absolute. It cannot be violated for any reason at any length scale, no matter how small, even in regimes where the underlying equations have never been determined. This belief may be correct, but it is an enormous speculative leap. One can imagine moon people applying similar reasoning and chastising their brightest students for asking what the earth was made of on the ground that its roundness made the question moot. This would clearly be an injustice, since the earth is not absolutely round but only approximately so. On length scales smaller than the naked eye can easily discern from the moon, there are troublesome little details such as the Grand Canyon, the Pamir, Aconcagua, and Kilimanjaro. Advances in observation technology would eventually vindicate the students, at least the ones who remained defiant. It would be discovered that the earth is not perfectly round, and moreover is approximately round for the *reason* that the rocks from which it is made become plastic at the high pressures found underground, so that large objects on the surface slowly sink.

Despite its having become embedded in the discipline, the idea of absolute symmetry makes no sense. Symmetries are caused *by* things, not the cause *of* things. If relativity is *always* true, then there has to be an underlying reason. Attempts to evade this problem inevitably result in contradictions. Thus if we try to write down relativistic equations describing the spectroscopy of the vacuum, we discover that the equations are mathematical nonsense unless either relativity or gauge invariance, an equally important symmetry, is postulated to fail at extremely short distances. No workable fix to this problem has ever been discovered. String theory, originally invented for this purpose, has not succeeded. In addition to its legendary appetite for higher dimensions, it also has problems at short length scales, albeit more subtle ones, and has never been shown to evolve into the standard model at long length scales, as required for compatibility with experiment.

Thus the innocent observation that the vacuum of space is empty is not innocent at all, but is instead compelling evidence that light and gravity are linked and probably both collective in nature. Real light, like real quantum-mechanical sound, differs from its idealized Newtonian counterpart in containing energy even when it is stone cold. According to the principle of relativity, this energy should have generated mass, and this, in turn, should have generated gravity. We have no idea why it does not, so we deal with the problem the way a government might, namely by simply declaring empty space not to gravitate. In chutzpah, this ranks with the famous case of the Indiana state legislature passing a law declaring π to have the value three.[11] It also demonstrates the severity of the problem, for one does not resort to such desperate measures when there are reasonable alternatives. The desire to explain away the gravity paradox microscopically is also the motivation for the invention of supersymmetry, a mathematical construction that assigns a special complementary partner to every known elementary particle.[12] Were a superpartner ever discovered in nature, the hope for a reductionist explanation for the emptiness of space might be rekindled, but this has not happened, at least not yet.

If Einstein were alive today, he would be horrified at this state of affairs. He would upbraid the profession for allowing this mess to develop and fly into a blind rage over the transformation of his beautiful creations into ideologies and the resulting proliferation of logical inconsistencies. Einstein was an artist and a scholar but above all a revolutionary. His approach to physics might be summarized as hypothesizing minimally, never arguing with experiment, demanding total logical consistency, and mistrusting unsubstantiated beliefs. The unsubstantiated belief of his day was ether, or more precisely the naive version of ether that preceded relativity. The unsubstantiated belief of our day is relativity itself. It would be perfectly in character for him to reexamine the facts, toss them over in his mind, and conclude that his beloved principle of relativity was not fundamental at all but emergent—a collective property of the matter constituting space-time that becomes increasingly exact at long length scales but fails at short ones. This is a different idea from his original one but something fully compatible with it logically, and even more exciting and potentially important. It would mean that the fabric of space-time was not simply the stage on which life played out but an organizational phenomenon, and that there might be something beyond.

Carnival of the Baubles

What of the future of this adventure? What will happen ultimately? We are going along guessing the laws; how many laws are we going to have to guess? I do not know. Some of my colleagues say that this fundamental aspect of our science will go on; but I think there will certainly not be perpetual novelty, say for a thousand years. This thing cannot keep on going so that we are always going to discover more and more new laws. If we do, it will become boring that there are so many levels one underneath the other. It seems to me that what can happen in the future is either that all the laws become known—this is, if you had enough laws you could compute consequences and they would always agree with experiment, which would be the end of the line—or it may happen that the experiments get harder and harder to make, more and more expensive, so you get 99.9 percent of the phenomena, but there is always some phenomenon which has just been discovered, which is very hard to measure, and which disagrees; and as soon as you have the explanation of that one there is always another one, and it gets slower and slower and more and more uninteresting. That is another way it may end. But I think it has to end in one way or another.

Richard P. Feynman

MANY THINGS IN NATURE ASSEMBLE THEMSELVES. IT IS fortunate that they do, for while we scientists advertise ourselves as exceedingly clever molecular architects, we are actually more like big Oklahoma tornadoes sweeping across the land, dealing out mayhem on a stupendous scale, and leaving a few interesting structures behind by accident. The pride one feels in making and observing natural self-assembly is not unlike the pride a father feels watching his son excel on the football field. It is indeed "my boy" out there, but the way he was actually made was amateurish, messy, and not at all a sure bet. There were attempts with exactly the same experimental conditions but different outcomes. The real reason Johnny is performing out there is that it is his nature to do so. I set the stage, but the essential things he did by himself.

I got a memorable lesson in self-assembly when I joined the faculty at Stanford many years ago. I had learned the rudiments of the subject previously but, like most physicists, was badly undereducated in the thing that counts, good old chemistry. That all changed when I got my new job. My duties included participating in a yearly cross-disciplinary technical review of the materials laboratory to which I was attached. This turned out to be quite sobering but also immensely valuable, for it exposed me to activities outside my immediate expertise, the first step toward new ideas for research.

About halfway through the daylong program of my first review I saw a presentation by our electron microscopist that knocked my socks off. This person's job was to take pictures of surfaces of materials—mostly inorganic crystals grown locally for other purposes—on a scale just below the resolution limit of ordinary light microscopes, a few tens of atoms to several thousands, which is also the scale on which much of the machinery of life operates. Her presentation was less a technical seminar than a *National Geographic* special on the Escalante Staircase or the Himalayan foothills of Tibet. She showed a sequence of the most astonishing topographies, no two of which were the same. First came layered plateaus punctuated with ragged canyons and peaks, their sharp vertical drops throwing the cliffs into shadow and exposing intricate systems of vaulted caves, followed by an archipelago of perfectly shaped pyramidal islands on an otherwise glassy smooth plain, a mad abstract Giza such as one might find in a Salvador Dali painting or *The Matrix*. Then came a forest of grotesque gargoyles on stems planted at the edges of lakelike dimples, thirsty cauliflower creatures from another planet descending on New England to seek out ponds. After that, a formidable and breathtaking mountain range with strange caps on the top resembling snow, as one might see from an airplane flying over Aspen or Katmandu. The show went on and on, and powerfully impressed upon me that I was surely in the presence of a genius, since no one in my line of work had ever discovered such wonders.

In science, however, as in anything else, missing an exciting investment opportunity can be a blessing in disguise. I had my hands full with academic responsibilities at the time of this review and thus was not able to drop everything, as I wanted to do, and make theories to explain these fabulous effects. A year went by. We had another review, and the electron microscopist again showed a stunning sequence of pictures, all different from the previous crop and all just as amazing. Then it hit me. This person had not discovered anything other than how to put samples into the machine. At the scale visible to the electron microscope, *every* surface looked interesting. Just as great talent would be required to take a dull photograph of southern Utah, so would great talent be required to take a dull electron micrograph of a surface. On this size scale, powerful and sophisticated principles of self-organization were at work in the inanimate world, many associated with the process of crystal growth and all quite unpredictable, notwithstanding our complete mastery of the underlying rules.

Seeing structures like these for the first time causes even a hard-boiled reductionist to pause and wonder whether they might be caused by some agency other than elementary quantum mechanics. It is one thing to explain ordered crystals of atoms with simple microscopic rules, but quite another to do so with complex lifelike structures and shapes, especially when one cannot deduce from first principles that these shapes should emerge. But this common and perfectly reasonable viewpoint is exactly backward. In a world with huge numbers of parts the unusual thing is not complexity but its absence. Simplicity in physics is an emergent phenomenon, not a mathematically self-evident state from which any deviation is a worrisome anomaly.

It is somewhat easier to explain and defend this assertion if you substitute the word *random* for *complex*. Thus you roll a die and the number three comes up at random. This statement means that you did not know ahead of time which face would come up, that it is something unpredictable, and that the degree of unpredictability is measured by the number of possible outcomes, in this case six. There is nothing random about the number three itself once it has been selected. It makes no sense for any particular die face to be "random." Similarly, it makes no sense for an isolated shape to be "complex." Only the selection of one shape out of many, a physical process, can be complex. When we say a shape is complex we really mean that the physical process by which it formed is unstable and with a slight nudge could have generated one of many different shapes. Similarly, we say a shape is simple if it is guaranteed to be formed by a physical process the same way every time, even when nudged fairly violently.

Once you understand that simplicity in nature is the exception, rather than the rule, it becomes easy to imagine that lifelike patterns might emerge if the microscopic circumstances were suitable. It is not possible to *prove* that they emerge, but it is possible to prove that their emergence is reasonable and does not violate common sense.

It becomes easy to imagine that lifelike patterns might emerge.

One does so by means of complexity theory, a branch of mathematics born in the 1970s that subsumes the topics of chaos, fractals, and cellular automata.[1] The strategy of complexity theory is to so simplify and abstract the equations of motion of matter that they can be solved reliably by computer. This abstraction, however, is a pact with the devil, since the resulting equations so grotesquely distort things that you no longer have a faithful representation of nature. The value of complexity theory is thus limited to showing that emergence of complex patterns is reasonable. It cannot supply predictive models of any natural phenomenon, and it is certainly not a fundamentally new way of thinking.[2]

A simple example of such a model is the mountain range fractal.[3] A computerized map grid is refined again and again, each time assigning a fictitious height to the new grid point that is the average of the heights of the adjacent old ones plus a random increment that becomes smaller and smaller as the refinement proceeds. The heights thus generated simulate the appearance of real mountain ranges so effectively that they are often used in movies to generate backdrops, as are their close relatives the fractal cloud, the fractal coastline, and the fractal vegetable (broccoli). The physical process being emulated by the mountain fractal is presumably aggregation, a process of surface growth by which an atom diffusing in from above sticks to the first place it hits, thus encouraging large structures to grow at the expense of small ones by shadowing them out. The large literature on diffusion-limited aggregation includes beautiful computer-generated patterns resembling the leafy ice crystals that sometimes form on windowpanes in winter.[4]

Another complexity model—legendary on account of being one of the first discovered—is John Conway's *Life*, a cellular automaton originally popularized by Martin Gardner's "Mathematical Games" column in *Scientific American*.[5] *Life* consists of a checkerboard with tokens that are removed from the board (death) or added to the board (birth) at each tick of an imaginary clock according to the following two rules:

1. A token dies unless two or three of its eight neighboring sites are occupied by other tokens.
2. A token is born on an empty site if exactly three of the eight neighboring sites are occupied by other tokens.

The tokens of *Life* generate patterns that resemble a whole range of natural phenomena, from solid crystals to little living creatures, which are given whimsical names by the large community of hobbyists who study them. Thus one can have stable space-filling crystalline patterns such as chicken wire and onion rings, small isolated molecular patterns such as rabbits and cows, cycling patterns such as blinkers and toads, rectilinearly moving patterns such as puffers and dragons, patterns that interfere with other patterns such as reflectors and eaters, and a whole zoo of complex higher organisms such as pedestals, fumaroles, rakes, cup hooks, beehives, replicators, volcanoes, aircraft carriers, and French kisses.

Both physical self-organization and the automata that emulate it are interesting. Exactly why is often hard to pinpoint, but two dry, overly intellectual explanations are particularly liked by governments and may be found in unhappy abundance in technical reports and grant proposals. One is that we are curious to understand how life could emerge from atomic minutiae—how one could mix a few chemicals together and, presto, out comes a puppy who loves you. The other is that we dream to engineer new kinds of useful gadgets and products for practical use, such as early warning sensors for obnoxious smells or machines that render leftover banana peels into gasoline. The argument sharpens when one combines the two, as in seeking gadgets that also emulate life or have health implications, like self-assembling robots, a cancer cure, or new limbs for amputees.

The *real* source of our interest, of course, is neither of these things but simply our inborn addiction to baubles. All of us have a powerful instinct to collect things that are "interesting" even when they are useless. This effect is what enables souvenir shops in Antibes and Sausalito to make profits selling polished bits of rock, even though one can find the same bits of rock on the beach, and even though it takes great talent to find a rock that does not become pretty when polished. It is also why so many of us have personal libraries of books we have not read, great boxes of ancient photographs of Aunt Marge at the Grand Canyon we never look at, and garages so full of junk that the car will not go in. It is the reason Imelda Marcos had all those shoes. It is also the entire business proposition of that bizarre new global phenomenon, the Giant Christmas Store: three stories stuffed to the ceiling with twinkly lights, dolls, little robotic men chopping wood, Black Forest plastic fir trees frocked with plastic snow, bins of little rocking horses, little chairs, little saxophones, little English beefeaters, little sheep, little grand pianos, little red glass balls, big red glass balls, blue glass balls, golden glass balls, garish Russian egg glass balls, glass balls with miniature electric train sets in them, the Crèche Room, the Music Box Room, the Angel Room, the Cuckoo Clock Room, and the cash register where all major credit cards are accepted, while "Silent Night" plays in an endless loop in the background even in July. I realized this phenomenon had spun completely out of control when I was in Japan last November and spotted Christmas trees in hotel lobbies and heard carols piped into elevators. Lest I be accused of unfairly targeting the enterprising Christians, let me also mention the big stack of cans of Air from the Holy Land in the Tel Aviv airport, and all those Arab shops lining the Via Dolorosa selling hookahs, useful brass pitchers, even more useful brass candlesticks, Crusader chess sets made in Taiwan, brightly colored Palestinian calendars, and crucifixes of all shapes and sizes along the way to the Church of the Holy Sepulcher.

The structures displayed by my colleague the electron microscopist are prototypical of what I call *nanobaubles*, fascinating and beautiful structures that develop spontaneously at small scales but have no known use except as entertainment. The size scale of the micrographs stretches all the way up to several thousand atoms, so these particular examples would be more aptly termed microbaubles, but I prefer the prefix nano because it is more general. Like the words *xerox* and *kleenex*, it has become generic and effectively a synonym for "very small," so that a nanobauble is actually just a tiny bauble.

My intention in coining this word, of course, is to satirize nanotech—the new technology of controlling matter on this length scale that will ostensibly lead us to a greater tomorrow. The need for such satire is not immediately obvious, for there is no argument that new organizational law is emerging at the nanoscale, that this law is potentially relevant to life, and that important discoveries are waiting to be made. The need is real, however, and clarifies after one has sat through enough presentations with dazzling pictures that never repeat, investigations that never come to closure, and arguments that never seem to hit home. The nanoscale traps you the same way the world wide web does when you log on and google for "mortgage interest rates." You get too much. Instead of a simple answer to a simple question you get a mighty river of sales pitches—page after page of brightly colored blinking banners promising much but delivering little. I once saw a television show in which the actor Tony Randall made a joke about being nibbled to death by ducks. That's what it's like. While our knowledge of the nanoscale is exploding almost incomprehensibly at the moment, nearly all of it is deeply unimportant. Predicting great new technologies from this situation is like predicting lasers from the existence of Christmas ornaments.

Even the list of industrially significant nanotech accomplishments turns out on close inspection to be dominated by inspirational but willful little nanobaubles. The nanotube, a small cigar-shaped structure a few atoms across made of pure carbon, appears to be a counterexample because of its many potential uses, but this appearance is incorrect.[6] Many nanotube uses, such as conducting additives to plastics, rely on chemistry and can be accomplished by other means, and applications such as nanotube-powered microsubmarines like those in Isaac Asimov's *Fantastic Voyage* are science fiction.[7] Nanopeapods—nanotubes with smaller molecules stuffed into them—are certainly nanobaubles,[8] as are the hexagonally packed structures of nanotubes called nanoropes.[9] In the specific case of semiconductor nanocrystals, objects much in the news lately because they have fluorescent properties similar to those of organic dyes (as do any chunks of semiconductor, actually), the various shapes are similar to the creatures generated by Conway's *Life* and have been given similarly creative names by their discoverers: rods, teardrops, arrowheads, tetrapods, branched tetrapods, and horns.[10]

How otherwise coldly logical people could fixate on such manifestly unimportant matters is a fascinating question—one ultimately answered, in my view, by the seductive power of reductionist belief. The idea that nanoscale objects ought to be controllable is so compelling it blinds a person to the overwhelming evidence that they cannot be. The idea also pervades the language we use to describe nanobaubles, which draws heavily on physical analogy with macroscopic things as a way of making them tangible. However, nanostructures are not macroscopic things, as becomes obvious when you pare away the rhetoric and computer graphics and describe actual experiments. Nanotubes, for example, are made not by adding carbon atoms one at a time, as one would with Tinkertoys or a knitting project, but by chemically separating the soot generated by a violent laser blast on a carbon target or a fiery carbon arc. Semiconductor nanocrystals are made not by patterning and lithography but by powerful electro-chemical etching with hydrofluoric acid in the presence of light,[11] or by grinding up conventional crystals and then rapidly injecting the powder into hot detergent. The list goes on and on. As with the sur-face preparations I encountered as a young professor, what is actually creating these objects is a higher-level rule of organization. One ac-tually controls not their blueprint but a temperature, a flow rate, a substrate orientation, or some other chemical condition.

Ironically, the illusion is reinforced by the wonderful modern measurement tools that allegedly overcome all these fundamental limitations by pure technical machismo. The trick to seeing through the deception is understanding how such tools work. An electron mi-croscope or scanning force microscope image of a nanobauble, for example, always begins by immobilizing it on a massive stage, caus-ing it to inherit integrity of mass from the apparatus. With the object thus immobilized you can collect information about it at your leisure, slowly building up a crisp image over time. Without immobi-lization you would have to take the picture fast, which would require radiation intensities that would fry the sample. (This very scenario is actually being discussed right now in the context of new accelerator-based X-ray sources, the hope being that you might get some infor-mation out before the sample blows up.) A corollary of this fact is that it's impossible to take pictures of nanobaubles while they are forming and thus impossible to falsify theories of why they exist. Even humble X-ray structure analysis of proteins exploits crystallization of the protein—an emergent process—as its first step. Thus, as a practical matter, all nanoscale measurements are clever exploitations of some emergent collective phenomenon, and thus all deliver artificial and highly manipulated representations of the ostensibly understood thing.

The mismatch between what you can "see" and what you can directly affect is unhappily reminiscent of some familiar aspects of medicine. I had an uncle, a neurosurgeon, who once invited me down to his hospital to view magnetic resonance images of brains. This came about because of a dinner conversation in which he asked me what I thought about such imaging, to which I answered, in typically arrogant physics student fashion, that it must be impossible. I had not understood the trick of varying the magnetic field strength from place to place in the measurement chamber, or that commercial products using this trick already existed. He was so amused by my response that he took time out from his busy schedule to show me his collection, which included not only interesting anatomical stuff but also grisly pictures of aggressively invasive tumors. He then sighed and admitted that diagnostic technology had greatly outpaced therapeutic ability, and that these people had, in fact, all died. The mismatch struck me as odd at the time, but in hindsight I realize that I was simply looking at neurosurgeon baubles.

Because of the physical circumstances here on earth—the temperature, the time scale of night and day, the chemical environment, and so forth—the most numerous examples of self-organization come from chemistry and involve aggregation of atoms, as opposed to some other kind of particle, into structures. We also know of examples from nucleons only, notably atomic nuclei themselves and the rules for isotope stability, and of electrons only, such as mesoscopic magnetism and Wigner crystallization,[12] but these are abstruse and require sophisticated machinery to detect. Thus while one can imagine lifelike behavior emerging in contexts other than those of ordinary chemistry, the experiments required to support these ideas are impractically expensive at the moment. An interesting consequence is that many chemists consider self-organization to be their exclusive purview and the practical dividing line between their own discipline and physics. This possessiveness sometimes has amusing consequences. I once sat next to Arthur Kornberg, the discoverer of the genetic replication enzyme DNA polymerase, at a dinner party. I was having a terrific conversation with him about the machinery of life when I made the mistake of opining that the whole thing had evolved into a wonderful physics problem. He then stopped, patiently explained to me that there was a lot of chemistry operating, and changed the subject. The poor fellow had heard it all before and wanted no part in maundering discussions about mechanical principles that one could not measure and had no bearing on experimental outcomes. I learned from this experience not to use the word *physics* in serious conversations with biochemists, especially those with medical training.

The conflict between physicists and chemists over who better understands emergent self-organization has its roots in an important and decidedly unscientific aspect of human psychology: To most of us, understanding a thing is synonymous with controlling it. For example, not understanding my kids means really that I cannot make them do what I want. Not understanding my car means that is uses more gas than I want it to, or burns oil, or will not start. You often hear people say: I do not understand this cable bill; I do not understand the government; I do not understand the opposite sex. You never hear them say: I do not understand my toilet; I do not understand my garden hose; I do not understand this celery. From a chemist's perspective, understanding a thing usually means making it and observing it, preferably before anyone else does. From a physicist's perspective, understanding a thing means categorizing it, making absolutely sure that this categorization is correct, and relating it to other similar things. Wolfgang Pauli's idea of "not even wrong" is central to physics but a total non sequitur in chemistry. Thus, on the matter of understanding there is total misunderstanding, one of these disciplines being from Mars, the other from Venus.

Unfortunately, while scientists bicker over who is the greater master of the universe, the nanobaubles are invading, multiplying with abandon, and taking over. Their nefarious plan is to change the rules by which the game is played: the more baubles one finds and the harder one works to enumerate their properties and make detailed genealogies of the pedigrees, the more blinded one becomes to the forest for the trees. Nanobaubles, it turns out, are from neither Mars nor Venus but from outer space.

What we're actually experiencing, of course, is not invasion of space creatures but a scientific paradigm shift—a large-scale reorganization in how we think forced upon us by events. It is plain to anyone not close to the problem that the carnival of the baubles represents something new in the history of human interactions with nature, and that turning it into science will require an invention—a social structure that combines parts of old disciplines into something adept at extracting the greater whole from the sum of the parts. It is also plain that this has not yet happened.

Partly as a result of these institutional inadequacies the situation in nanoscale physics and the interface with biology at the moment resembles less a happy academic retreat than a Western in which free-range cattlemen are at war with the sodbusters and their fences, while the railroad quietly buys up all the land that counts and bribes the legislature to look the other way. The similarity is not an accident, for this size scale, and the principles of self-organization operating there, is where the frontier of modern science resides. It is an exhilarating place to be, and a natural home for many of us, but it is not for sissies. As was the case with the Wild West, the rules of personal behavior in this realm are somewhat ill-defined because there is no government yet. Folks are busy staking out claims first and asking questions later, conducting their businesses and their lives as best they can given the opportunistic social chaos. Lots of money is sloshing around, and great fortunes are being made and lost in a few hands of poker or a gun duel in the dirty main street of town. There are also big-time land and mining swindles, and lots of snake oil and patent medicine being sold. But now, as then, the allure of traveling in such a wild and lawless place is the ever-present possibility of making a serendipitous discovery of great importance.

Faced with such glaring and persistent failures to exploit opportunity, it is sometimes hard to keep the faith that discovery is inevitable, at least in the context of institutions we have now. Before succumbing to temptation and quitting, however, it is wise to recall how intractable previous generations found the problems facing them, and how they courageously followed clues dropped by nature to breakthrough and solution. The miracle of color in nature pointed to chemical principles that eventually led to the invention of aniline dyes. The miracle of rectification in rocks pointed to the principles of semiconduction that led to the invention of the transistor. In each of these cases, moving forward required the invention of entire thought processes, practices not recognized as lacking until long after they had been invented. Today we are contemplating the miracle of life and the principles of organization at the nanoscale to which it points. It is conceivable that this problem will be impossible to solve, but I don't think so. We have the same important piece of information about this problem as we had for dyes, semiconductors, and all the other technical marvels that have now made their way into the economy and become woven into our lives: nature already did it. Admittedly, it had a vast stretch of geologic time over which to do its research, but then that is also true for the other things.

I once went out for several days in the high country north of Yosemite with one of my sons and two friends. Accommodating everyone's schedules required us to go in August, something inherently problematical on account of water. It does not rain much in these mountains in the summer, and late in the season the last of the snow has melted, the streams have dried up, and travel must be planned around the few available lakes. It is also hot. Above the tree line there is only barren rock and sandy glacial rubble, and large stretches are little more than nasty deserts, even though the altitude is very high.

On the third day out we had to cross a particularly long and brutal stretch of this desert, and just barely had enough daylight to reach a small lake marked "unsuitable for camping" on the map. There was no practical alternative to shooting for this lake, unfortunately, so we decided to ignore the warning and endure whatever unpleasantness might be in store for us there for one short night. What a mistake. After exhausting our water and strength trekking across treeless waste all afternoon, we descended to this lake and found it to be a shallow, reedy, brackish sump infested with mosquitoes and difficult to access at all on account of the broadness and muddiness of the bank, which was festooned with deer and cattle hoofprints. Moreover, this lake was surrounded on all sides by moonscape, the only redeeming feature of which was an unobstructed view of the Brown Bear Pass, our escape route out the next morning.

I was too tired and dehydrated to even think about abandoning the plan and continuing over the pass that night, especially since we were not sure about water on the other side. I had been worried about this very possibility, actually, and had asked a horseman we met on the trail earlier in the day where the next water was. He said he thought there was some a couple of miles beyond the pass, but was not sure because he had not been there for several weeks. But the issue was moot, for the younger fellows simply could not stand this lake, felt strongly that we should move on, and eventually talked us into trying.

So, with grim determination we headed silently up the long slope, which eventually steepened into seemingly endless zigzags and switchbacks, and just managed to gain the top as the light failed. The other side of the pass was a long vertical descent through shadows and ankle-twisting talus chunks to a dry watercourse at the bottom of a box canyon. We had stumbled down to the bottom of the rock-fall and were just about to break out the flashlights for the treacherous trip down the canyon when I heard it: the faint but unmistakable sound of water. That guy had been wrong. There was a little spring at the bottom of the cliff hidden in the willows. We were saved.

I do not remember the rest of the evening in great detail because I was a bit delirious, no doubt from being careless about salt, but it worked out well. We built a small fire on a slab of granite, cooked up some indifferent freeze-dried stuff, retired to our beds, and fell instantly asleep. But I remember impressions: the scent of willow and sage, a jet-black sky rimmed with cliffs and ablaze with the glory of the Milky Way, the gentle murmur of the brook, and the occasional mournful whisper of the wind echoing quietly from the rock walls. There was a coyote howling way down the canyon too, but it eventually got tired and went away.

There are springs in the wilderness where people do not go, and refreshing water no one knows. But to find it you must strike out beyond the parts, study the land, level with yourself when you have misunderstood something, and trust Providence.

TWELVE

The Dark Side of Protection

Nature is wont to hide herself.

Heraclitus

Everyone with a weakness for escapism—which is everyone, for the most part—knows about the Dark Side of the Force immortalized in *Star Wars* movies. This great mythic archetype is the evil aspect of what Stoic philosophers called *natural order*, an overarching principle or substance they thought informed the universe. The Dark Side is always lurking out there to corrupt you. Strong people avoid its temptation, but weak ones, like Darth Vader, succumb. The Dark Side's devilry begins to matter in earnest when Darth and the other dark graduates get together and conspire. One of them, the evil Senator Palpatine, successfully becomes emperor of the galaxy by recruiting others to the Dark Side and manufacturing fictitious threats to peace and stability that he can handle "reluctantly" if given supreme power by the legislature. In fear for their safety and yearning for protection, the people grant the senator these powers, only to see him consolidate them into a cruel and oppressive dictatorship.

Not only governments and annoying private-sector versions of them, such as the Mafia, but also nature itself provides protection through laws insensitive to destabilizing outside influences.[1] Protection generates exactness and reliability in the physical world just as it does in the human world, but its physical versions have the advantage of being primitive, so that one can unambiguously identify them as spontaneous self-organizational phenomena involving no intelligence other than the principle of organization itself. The universal properties of ordering of rigid bodies, the flow of superfluids, and even the emptiness of space are among the many concrete, well-documented examples of this effect.[2] The insensitivity of material rigidity to an atom out of place is no different from the insensitivity of an election outcome to an individual political opinion out of place—which my own somehow always manages to be. In the end, protection overcomes imperfection with the single-minded efficiency of a doting mother watching a parade march by and observing, "Oh look! Everyone is out of step except my Johnny!"

Like their human counterparts, however, institutions of protection in nature have a dark side—their tendency to circumscribe one's options by obscuring ultimate causes. The elastic rigidity of the solid state, for example, a powerful law that makes possible the engineering of reliable structures, hides the existence of atoms, because the elastic properties are universal consequences of ordering and would be the same if the solids were made of something else. Proving the existence of atoms is fundamentally impossible without a measurement technology, such as X-ray scattering, capable of escaping the protection. Ignorance of atoms does not matter if you are manufacturing cars or skyscrapers, but it matters a great deal if you are trying to make computers or television sets. The haphazard nature of technological advance is thus arguably a consequence of the dark side of protection, as is the perception that technologies are somehow "unnatural." The extreme case of this problem is the vacuum of space itself, which shows signs of universal protected behavior in the experiments we are presently able to do, and thus regulation by microscopic laws that we cannot know until our accelerator technologies improve.

The dark side of protection has many everyday analogies. McDonald's, Starbucks, and Kentucky Fried Chicken, for example, are useful precisely because their products are stable and reliable, so that you know ahead of time what you are getting. At the same time, however, patronizing them exclusively closes out the possibility of ever discovering a restaurant that innovates. This is why independent-minded people so dislike these companies—even while owning stock in them and relying on them in a pinch when certainty matters. Stock or no stock, however, I will not eat franchise frozen yogurt even when desperate in an airport, and I warn people who lead dissolute lives that their reward in hell will be to sit chained to a table for all eternity with a menu offering only chicken Caesar salad. Fortunately for all of us, there is less protection in France. Somewhat grimmer humor along these lines may be heard from Russians who lived through the Soviet days, or Harlan Ellison's *A Boy and His Dog*, a deliciously black satire in which post-World-War-III society protects its "way of life" by banning novelty entirely.[3] Ellison's hero is a sex-crazed misogynist who escapes the protection and returns to the surface of the earth to save his starving telepathic dog by feeding him his cloying girlfriend. My wife does not like this story.

All well-documented cases of protection in physics are characterized by invariance of scale.[4] This idea is illustrated by the story of the incompetent director who wants to make a movie of an organ pipe sounding. This is obviously not a big money-maker, but he is an avant-garde director and believes that this film will be the ultimate Zen cinematic experience. After a few minutes of filming he decides that the quality is not good enough, so he yells "cut" and regroups. Technicians are called in to make a pipe with all the dimensions doubled—which of course sounds with a lower tone—and the cameraman is instructed to back up, so that the enlarged pipe again fills the field of view. He then begins filming again—until he realizes his mistake. In a rage he jams the developed film into the projector, flips a switch to make it run twice as fast, and confirms that, sure enough, the image and sound are exactly the same as they were before. The improvements changed nothing. The reason is that the laws of hydrodynamics responsible for the sound of the organ pipe are scale-invariant. The pipe's observed behavior remains the same if the sample size is doubled, followed by a corresponding doubling of the scales for measurement of distance and time. This process is called *renormalization*, and it is the traditional conceptual basis for discussing protection in physics.[5]

Renormalizability is fundamentally lopsided. In the case of the organ pipe, for example, one can scale to larger and larger sizes forever without breakdown of the renormalization rule, but scaling in the opposite direction, to smaller sizes, works only down to the size of the atoms, at which point the laws of hydrodynamics fail. It is actually more revealing to imagine this experiment in reverse—starting from a small sample and scaling up. One finds that corrections to hydrodynamics such as atomic graininess, nonlinear viscosity rules, dependence of the flows on internal factors other than pressure, and so forth get smaller and smaller with each size change, causing the phenomenon of hydrodynamics to "emerge" in the limit of large sample size. That's the good news. The bad news is that there are other possibilities. If the average number of atoms per unit volume had been slightly higher, the universalities of crystalline solids would have emerged under renormalization instead of those of fluids. One might say that small samples contain elements of all their possible phases—just as a baby contains all the elements of various kinds of adulthood—and that the system's identity as one phase or the other develops only after some properties are pruned away and others enhanced through growth.

The technical term for diminution of some physical property, such as shear strength in fluids, under renormalization is *irrelevance*. Thus, in a fluid, corrections to hydrodynamics—indeed most properties of a collection of atoms one could imagine measuring—are irrelevant, as are the corrections to elastic rigidity when the system is solid. Unfortunately, irrelevance is also a strong contender for the dumbest choice for a technical term ever made. It confuses everyone, including professional physicists, on account of having multiple meanings. I could go on and on about the practice of rewarding scientists for inventing things other people can't understand, but will constrain myself and note only that one of the easiest ways to do that is to assign a new meaning to a commonly used word. You chat along, casually drop this word and, presto, anyone listening is hopelessly confused. The trick to deciphering the code is to realize that there are two versions of the word "irrelevant." One means "not germane" and applies to lots of things other than physics. The other means "doomed by principles of emergence to be unmeasurably small" and applies only to certain physical things.

The emergence of conventional principles of protection acquires an interesting twist when the system is balanced at a phase transition, so that it has trouble deciding how to organize itself. Then it can happen that everything is irrelevant except one characteristic quantity that grows without bound as the sample size increases, such as the amount of magnetism in a magnetic material. This *relevant* quantity ultimately decides which phase the system is in. In the case of magnetism, for example, the growth is negative if the temperature exceeds a certain value, causing the magnetism above this temperature to disappear. Being magnetic is all or nothing. There can also be quantities that neither grow nor diminish, but these so-called marginal variables characterize a special kind of stillborn phase transition that occurs rarely (i.e., never) in nature. The situation resembles a tug-of-war between well-balanced teams. When the game begins, the teams wrestle back and forth—first one gaining the advantage, then the other—exhibiting the universal characteristics of a tug-of-war, all other aspects of life having become irrelevant. At last, one team accelerates the rope faster and faster to its side, and the other team loses control and gets dragged catastrophically into the mud. That there will be a winner of this contest is certain, but the time it takes for the winner to emerge is not. In principle, the contest could be drawn out for an arbitrarily long time if the two teams were arbitrarily well balanced. In practice, the balance makes them highly susceptible to external influences, such as rainstorms or heckling from bystanders, which then decide the outcome rather than the natural superiority of one team over the other. This effect also occurs in balanced elections, which is why such elections mean very little.

Balanced protection occurs commonly in nature, but less so than one might anticipate because most phase transitions, like the evaporation of water, have a latent heat that forces the phases to coexist. Water is nicely in phase balance on a hot, humid day, when some is in the air and the rest in lakes and ponds. This balance is precisely what makes these days so uncomfortable, for it prevents the water in one's body from cooling it by evaporation. But if the water is placed under pressure, the heat required to turn the liquid into vapor can be made to diminish and then disappear altogether, thus obliterating the difference between the liquid and vapor. When it just barely vanishes one gets a true balance effect called *critical opalescence*, in which the fluid becomes milky and opaque.[6] This effect is something like fog, but vastly more interesting because it lacks scale. The droplet size in real fog is determined by environmental factors, such as dust and microscopic bits of sea salt in the air, and could just as easily have been extremely large—the extreme case being a nearby lake. But under pressure the schizophrenia of the fluid is maximized, and its foglike behavior exists on all scales simultaneously. While this effect is wonderfully entertaining to see, its practical use is restricted to steam turbine design, which exploits this special property of its working fluid to maximize fuel efficiency.

Balance universalities and relevance associated with phase transitions in nature cause two physical effects I call the Dark Corollaries. The melodramatic overtones are intentional, for these effects are insidious, destructive, and thoroughly evil, at least from the perspective of anyone concerned with differentiating what is true from what is not.

The first Dark Corollary I call the Deceitful Turkey effect. The name comes from a Mark Twain sketch in which he describes hunting a turkey that is feigning injury so as to lead the hunter away from her nest.[7] He manages to just miss her again and again, and only after having been led miles astray does he realize he was not actually close at all but merely taken in. The Deceitful Turkey effect in physics is similar. While stable protection prevents us from determining underlying microscopic rules, *un*stable protection tricks us into thinking we have found them when we actually have not. This, in turn, makes the empirical case that the effect exists rather lengthy, since the corresponding experimental literature is confused. A parable is clearer. Returning to the tug-of-war, let us suppose we are trying to find the "first cause" of tug-of-war games by conducting observations at shorter and shorter time scales. Let us further suppose that the game is well balanced, so that the scale for a decisive outcome is extremely long, thus making the realm of indecisiveness easy to reach experimentally. Then, conducting the experimental observations, we discover a broad range of times over which tug-of-war universality is manifested—the teams remain balanced independently of the participants, the nature of the rope, the slipperiness of the ground, and so forth—and moreover, this behavior is legitimately observed to be the precursor or progenitor of the decision and thus the "underlying" law. The universality of the behavior also permits a simple mathematical description of it, and, by implication, a simple mathematical description of the descent to a final outcome. We think we have found the simple ultimate cause of the tug-of-war, but what we have actually found is intermediate protected behavior masquerading as the ultimate cause! Our theory of the descent is correct, elegant, mathematically rigorous, and totally meaningless. We've been duped! In this case, however, the deceiver is not a crook or colleague (or a turkey) but nature itself.

The second Dark Corollary I call the Barrier of Relevance. Suppose by some miracle one was able to discover the true underlying mathematical description of a thing, whatever it was, and took as one's task to solve the equations and predict the protected behavior that they imply. One would have to make approximations, of course, and in a stably protected situation, the small errors implicit in these approximations would be irrelevant in the technical sense that they would heal as one scaled up to larger and larger sample sizes. But in an unstable situation, relevant mistakes grow without bound. Rather than healing one's errors, the physical behavior amplifies them, causing one's prediction to become less and less reliable as the sample size grows. This effect is conceptually the same thing as "sensitive dependence on initial conditions" in chaos theory, but differs from it in pertaining to evolution in scale rather than evolution in time. As in chaos theory, a very small error in the procedure for solving the equations can metastasize into a gigantic error in the final result—large enough to make the result qualitatively wrong. This kind of universality destroys predictive power. Even if you had the right underlying equations, they would not be of any use for predicting the behavior you actually care about, because you could not solve them sufficiently accurately to make such predictions. This, in turn, makes them unfalsifiable.[8] If you cannot predict certain experiments reliably, then you also cannot use these experiments to determine whether the theory is correct. The system has spontaneously generated a fundamental barrier to knowledge, an epistemological brick wall. Yet inside a given phase, the macroscopic properties are quite predictable. It is like dating. The underlying impulses are simple and easy to understand, the end result is one of a handful of universal phenomena, but what comes in between is complex and highly unpredictable.

A textbook example of Dark Corollaries at work is the correlated-electron effect.[9] The name itself is actually a reductionist misnomer, for "correlation" in quantum mechanics just means "entanglement," something that electrons exhibit always—not just some of the time. Calling electrons correlated is like calling bodies of water wet. The correlated-electron effect is actually a set of behaviors in solids that do not fit the traditional categories of simple metal, insulator, ferromagnet, and so forth, but instead seem somehow stuck in between. They occur mostly in metallic oxides (V_2O_3), but also in certain intermetallic compounds ($CeCu_2Si_2$), alloys (UBe_{13}), and many organic substances (charge transfer salts). In addition to their notorious classification difficulties these materials have a long list of suspicious properties such as hypersensitivity to atomic imperfection, ordered phases that come and go depending on sample preparation method, and spectroscopic properties that do not reproduce, suggesting proximity to one or more poorly characterized phase transitions. However, the name "correlated" implies something else: the usual approximate techniques for describing entangled matter do not work in these materials for some reason, and that's the problem. In other words, the behavior is strange because you cannot calculate it—as opposed to being difficult to calculate because it is strange.

You might think that such a basic issue would have been easily resolved by experiment, but this is not the case. Year after year, different research groups would obtain different answers from the same experimental measurements, often even different from their own results a few months before, and year after year they would defend the integrity of their work by accusing others of incompetence. Theorists would then "analyze" these results by picking them over, selecting the ones they liked, and pronouncing these to be morally superior by virtue of supporting their case, whatever it was. The underlying equations were perfectly well known, of course, and perfectly amenable to approximate solution in less controversial contexts, but they could not be solved with sufficient accuracy to predict what should happen in these experiments. This had the lovely added bonus of making the theoretical intransigence itself controversial, since one could always argue that the other person's calculations were flawed. Thus in the fifty years since people began working on the correlated-electron effect in earnest, no progress has been made in clarifying the thing that counts: what the effect is.

In light of hindsight, such scurrilous behavior is a symptom of asking extremely bright people to do impossible things, and thus effectively proof that a Barrier of Relevance is at work. It is fundamentally impossible calculate reliably through such a barrier, even with the largest computers, hence the variety and mutual incompatibility of the theories. It is fundamentally impossible to stabilize the experiments against a material's nuances, hence the failure of the experiments to be reproducible. It is fundamentally impossible to falsify theories—in this case approximation schemes built into the relevant computer programs—hence the political nature of the arguments about them. The theories were all Deceitful Turkeys— ideas that might be shown to be valid some day but for now are just out of reach.

Scientists trapped by the Dark Corollaries often intuitively understand that something is wrong but cannot put their finger on exactly what it is, and so make jokes. The following story is told by researchers of high-temperature superconductors, correlated-electron materials of special historical importance by virtue of being so thoroughly (and unprofitably) studied. A small country experiences a coup d'état, and the new government sets out to execute all the members of the old cabinet. Two of them are hauled in front of the new dictator to receive their sentences. He offers each a last request. The first one says, "Well, before I worked in government I was a physics professor. My last request is that I would like to give a lecture, to be attended by every physicist in the country, on my theory of high-temperature superconductivity." The second one says, "I am also a physicist. Please kill me before he gives his lecture."

Another nice example of the Dark Corollaries' mischief, one that finished off many physicists when I was a student, is the notorious phenomenon of silicon surface reconstruction. It was discovered back in the 1950s that the atoms on the surface of a silicon crystal freshly cleaved in vacuum spontaneously move to generate ordering patterns. While the specific pattern one got depended on the cleaving method, annealing history, and so forth, the final and most stable pattern always had a repeat unit that was seven times longer than the distance between the atoms on the native surface and was squashed into a trapezoid. No one knew why the silicon did this, or even what the atomic rearrangement was, since the electron diffraction effects that revealed it could not determine the structure of the repeat unit sufficiently accurately. The great challenge at the time was to solve the equations of quantum mechanics by computer to figure out how the atoms moved to achieve this effect. How many man-hours of labor went into this problem I shudder to think, but they might as well have been sent into a black hole. It was simply too hard. All sorts of interesting patterns came out of these calculations, but none of them matched the experiments, a sure sign of Dark Corollaries at work. The structure was finally solved by Kunio Takayanagi, an experimentalist at the Tokyo Institute of Technology, using a new high-energy electron diffraction technique—after which lots of revisionist theories sprang up explaining why it had been obvious all along.[10] But this claim is untrue. To this day no one knows why the stable repeat length is seven, why it is skewed into a trapezoid, and why it is so stable—even though nature has no problem making the atoms order this way, over thousands of atomic spacings, every single time.

While the Dark Corollaries are best documented in materials science because of the huge mass of experimental information there, the place they matter most is cosmology.[11] It has been known since the 1950s that the vacuum of space is renormalizable—meaning that the elementary particles propagating through it and the forces between them obey the same kinds of scale-invariant equations you find emerging at phase transitions in ordinary matter. We also know that these things must be linked to space itself in some fundamental way because they do not generate gravity, hence the idea that space itself is renormalizable. Whether the renormalizability of the universe is generated by proximity to a phase transition is not known one way or the other, for one of its effects is to prevent you from inferring anything about short length scales from measurements made at long ones, just as happens with ordinary matter. Renormalizability is thus enshrined in textbooks as a property of space that just is—in keeping with the standard practice in science of postulating minimally. However, if renormalizability does not emerge, then it demands explanation, since it is miraculous, and a good rule of thumb in physics is that miraculous things have only one cause. Moreover, the vacuum is known to be near phase transitions. There are numerous experimental indications that the vacuum emerges in a hierarchy of phase transitions in which the various forces of nature differentiate from each other. The one of them associated with the distinction between electromagnetism and the weak nuclear force is central to modern cosmology, because the energy released when it occurs is the ostensible power source for inflation of the hypothesized brief period of rapid expansion following the big bang. If renormalizability of the vacuum is caused by proximity to phase transitions, then the search for an ultimate theory would be doomed on two counts: it would not predict anything even if you found it, and it could not be falsified.

The Dark Corollaries also have important and disturbing consequences for business and the economy. This is a difficult matter to discuss openly, since calling this or that activity deceptive or fraudulent is grounds for a lawsuit, so I will describe it allegorically. Any resemblance to real people or circumstances is the purest coincidence. Suppose I write a computer program that allegedly predicts something. I tell you the underlying equations—in other words, what the code ostensibly does—but do not reveal the method by which I solved them. I allege that solving these equations correctly is just a matter of being smart enough, and that poor souls such as yourself, who are not endowed with enough brains, just cannot do it. You are angered by this insult and go off to write a program that solves these equations your own way. But alas, after working many months to get it right, you not only fail to get my result, you get *different* results depending on how you execute the approximation scheme. You are sure you wrote your code correctly and now realize that I simply lied, since the equations are unstable. Indeed you begin to suspect that my ostensible prediction was actually an after-the-fact fit to the facts, a complete fabrication. The equations I revealed were not sufficient to describe what my program does, nor is it true that only those with special mental powers can solve them. No one can solve them! Proving this, however, is impossible for the same reason that they cannot be solved in the first place, nor can you check what I did to see whether it is right, because it is proprietary. So you are checkmated. The most you can do is write a paper or patent saying you have some "technology" that does something different from what mine does and has a different practical application.

We can put a positive spin on the situation by saying that unstable physical systems are important economically because they enable us to reveal the fundamentals of a thing without revealing the thing itself. This is fine for deceiving one's opponents and maybe gaining market share at their expense, but one must not kid oneself that it is anything more than what the character Yogurt in the Mel Brooks film *Spaceballs* calls "the soich for more money." Unfortunately, it can also destroy the lives of scientists, who think they are chasing a pot o' gold but are in fact chasing a rainbow.

One can imagine that I am none too popular saying things like this, but I do not care. It is better to be on target and hated than craven and beloved, and anyway, I have sacrificed plenty on the altar of irrelevance and thus know what I am talking about. But for those who are still not satisfied, I am selling little Dark Lord dolls in a likeness of myself, which they may purchase and then do with as they please. You pull a string and the doll squeaks out, "May the Schwartz be with you." It is adorable.

THIRTEEN

Principles of Life

You dehumanize a man as much by returning him to nature—by making him one with rocks, vegetation, and animals—as by turning him into a machine. Both the natural and the mechanical are the opposite of that which is uniquely human. Nature is a self-made machine, more perfectly automated than any automated machine. To create something in the image of nature is to create a machine, and it was by learning this inner working of nature that man became a builder of machines. It is obvious that when man domesticated animals and plants he acquired self-made machines for the production of food, power, and beauty.

Eric Hoffer

THERE IS NOTHING MORE HEARTWARMING THAN PONTIFICATIONS from computer executives about life. One hears a good deal less of it now that the dot-com bubble has burst, but it is still out there in people's brains, waiting for an opportunity to resurface and relieve you of your money. The less ambitious versions of computer imperialism, of course, never missed a beat, and the market still teems with time-devouring toys for grownups, while newspapers across the world run story after story about "technology," journalist-speak for computer programs. But the application of computers to life is in a

There is nothing more heartwarming than pontifications from computer executives.

class by itself. The outrageous chutzpah of computing experts on this subject reminds me of a line by the science fiction writer Robert Heinlein: If you're going to have a circus, you've *got* to have elephants.

Life is especially fun to talk about from a physical perspective because it is the most extreme case of the emergence of law. In fact, the entire idea of emergence was invented by biologists to explain why some aspects of living things—the rodlike shapes of some bacteria, for example, or the tendency of bunnies to run away from foxes—are stable and reproducible, while the microscopic laws of chemistry from which they descend are random and probabilistic. There are lots of examples of such things from intermediate-scale chemistry—gels, surface structure of crystals, and so forth—but the granddaddy of them all is the functioning of large organisms, such as people.

One of the common denominators of life is the powerful déjà vu experience. I had a big one recently in the middle of a molecular biology seminar. It was a massive PowerPoint presentation in which levels of six thousand types of messenger RNA went up and down (or not) over the cell cycle of yeast. This was, in addition to interminable, thoroughly exasperating, for although they are supposedly a window on the cell's basic regulatory machinery, no one knows why these measurements take on the values they do, what the crude correlations of one signal with the next imply, or indeed whether there is any useful information in these measurements at all.[1] At any rate, I was teleported back to a seminar I attended in the 1970s on color centers in silicon dioxide, watching optical absorption features go up and down (or not) in response to various kinds of violence done to the sample. The subject was different and the experimental technique much cruder, but the logic was exactly the same. Back then, the issue was not the machinery of life but chemical and structural defects in the oxide, which are highly detrimental to silicon microcircuits. They are easy to detect, fortunately, for they are highly efficient absorbers of light in otherwise transparent materials, this being why most rocks are colored rather than clear. They are also sources of spin resonance signals—the ability of the material, when placed in a magnetic field, to absorb energy at certain highly specific radio wavelengths. The objective of the study was to correlate the optical absorption properties with the spin resonance signals, and thus figure out which defect was causing which signal. But because there were too many defects to be isolated and studied individually, the strategy was to "perturb" the sample—for example, by cooking it for days in a kiln or placing it in a nuclear reactor overnight—and see what happened. The analogous experiment in biology would be to poison the yeast or starve it nearly to death. The hope was that one or two optical signals would grow at the same time as the corresponding spin signals, allowing one to ascribe the same defect to them. What resulted, of course, was utter pandemonium. Everything changed and correlated with everything else. It was like "perturbing" the sales floor of Bloomingdale's by announcing a storewide 90%-off sale. There were enormous effects, and the theorists naturally went wild, offering all sorts of plausible explanations that were consistent with the facts but as different from each other as night from day. But this diversity of opinion was, then as now, simply symptomatic of a poorly designed experiment—one incapable of answering the question.

Bad experiments are, unfortunately, endemic to higher-level science. The underlying reason is that teasing out details of how a complicated thing works is difficult and costly in time and labor, and thus money. The economic facts being what they are, it is almost always wiser to let someone else do the thankless work and to plan one's own program around cheap experiments with potentially high payoff. Ignoring these economic basics can be deadly, especially in business. If Boeing began worrying about why air molecules collectively generate hydrodynamics, it would clearly be time to divest Boeing. But in extreme cases, such as gene transcription, the thankless work may not get done by anyone, and the discipline is left with a logical hole to be filled in later, if ever. Some of the disconnect between microscopic law and sophisticated high-level behavior, especially in studies of life, is built into the way we do science.

The bad-experiment effect with which I am most familiar is not in biology but in nuclear weapons. Back when I was working at the Livermore Laboratory I would occasionally run across people who had worked for a long time with the nuclear design codes and had amusing stories to tell about them. These codes contain lots of stuff that I am not at liberty to discuss, but they are crudely analogous to the functioning of a cell, in that they are massively hierarchical: first this has to happen, then that, then that, then these two things combine in tight timing to do that, and so forth. These were all extremely good people, so the story always had gravity and staying power—and was always hilarious. It would revolve around some serious mistake in the person's field of expertise, which was different from one person to the next, and would be functionally uncorrectable because the designers did not believe that the problem mattered and would not countenance the requisite fix. Moreover, the mistake was never just a little detail but a violation of the second law of thermodynamics or the immaculate conception of energy—the sort of thing that elicits screaming laughter at the water cooler and inspirational references to Dilbert.

After I heard enough of these stories, however, I began to realize that this lunacy was not personal incompetence but a sociological phenomenon inherent in the discipline itself. What mattered was that the codes should guide one to the important outcome, the yield, not whether they did so logically. They had been adjusted to match the large-scale outcome of certain tests in the past, and would not function correctly (in other words, match these tests) if they were modified. The experiments to look inside the nuclear weapon to check the correctness of the theories implicit in the codes had never been performed, and probably never would be. Such experiments would be devilishly difficult to perform, for one thing, for it gets rather hot in there and one does not have much time to get the signal out before the measurement apparatus evaporates. The real reason, however, is that they would have to be done repeatedly with exquisite attention to detail, and would thus be extremely expensive. Fortunately (or unfortunately, depending on your perspective), there are wide margins for error in nuclear design due to the abundance of energy available. As in many engineering matters, the "truth" was determined by business needs, not by experimental reality, especially as defined by a bunch of academics. As long as the bombs exploded, the second law of thermodynamics could go hang. The joke, it seems, was on us.

The bad-experiment dynamic was also at work in the silicon dioxide defect problem. Behind the sophisticated academic agenda of categorizing all those defects lay the simple question of how to eliminate them in semiconductor manufacturing—which engineering houses eventually answered for themselves using good old Edisonian empiricism. The one exception is flash memory oxide, which stores signals in the defects and is thus made defective on purpose, the specific defects and techniques to generate them being trade secrets.[2]

The messenger RNA experiment in yeast is an especially important kind of bad experiment, however, because it demonstrates clearly that geneticists do not know what they are doing. The screams of outrage and other indignant responses to this assertion will fall on deaf ears: I know a terrible experiment when I see one. The symptoms are always the same. The measurements do not reproduce, they do not lend themselves to commonsense analysis, and they cannot be quantified. The argument that animate things are just fundamentally different from inanimate ones in this regard is false. There are plenty of highly quantifiable things in biology: the ribosomal genetic code, the fidelity of DNA replication, the crystal structures of proteins, the shapes of self-assembled virus parts, and even sophisticated behavior of higher organisms such as rats and people. The truth is that the control machinery for converting genes to life is not understood, and one of the key reasons is that such understanding would be prohibitively expensive to obtain.

That biotechnologists often do not know what they are doing is neither surprising nor an accident. Like the semiconductor physics of a previous era, biology has now evolved from science to highly profitable engineering. This distinction strikes most people as a mere label change, but it is actually a tectonic shift, for science and engineering differ in one central respect: in science, you gain power by telling people what you know; in engineering, you gain power by *preventing* people from knowing what you know. Chronic confusion and ignorance are the rule, rather than the exception, in engineering for the simple reason that everyone is withholding information from everyone else on intellectual property grounds. In the Silicon Valley, where I live, technical deception and bluffing are both commonplace and expected, and it is universally understood that admitting weaknesses in one's experimental investments, especially extremely expensive ones, would be economic suicide. The engineering value in biotech is not in understanding life but rather in designing drugs, inventing new health therapies, and creating new artificial organisms for agriculture. For these purposes, correct theories of regulatory processes are less important than rough, simple ideas that can motivate chemical manipulation. It has turned out that one can design protease inhibitors for control of AIDS,[3] trick stem cells into growing into replacement body parts,[4] and insert an alpha-carotene gene into rice[5] without understanding the regulatory machinery of cells at all. It is even possible to invent effective cancer therapies, despite the fact that cancer is fundamentally a malfunction of cell regulation, because the objective is to kill the cancer, not understand it. But beneath these stunning technical successes is the scientific loose end that the manipulators do not, in fact, know what they are doing.

I find it profoundly ironic that the very fallibility of science that motivated Mary Shelley to write *Frankenstein*—people's tendency to believe they understand things when they actually do not—should become mainstream and acceptable for financial reasons.[6] It brings to mind Oscar Wilde's observation that lack of money is the root of all evil. One can just imagine what Mrs. Shelley's novel would have been like had it been written today. Instead of an angst-ridden nerd from Geneva, Victor Frankenstein would have been an enterprising young spam profiteer from the Thomas Jefferson High School of Science and Technology in Alexandria, Virginia. Instead of traveling to Igolstadt to learn his highly creative surgical techniques, Victor would have flown to Boston to attend Harvard Medical School—after a four-year hiatus at Princeton investigating tennis and women. Instead of constructing the monster in secret he would have used political connections to obtain a stupendous grant from the National Institutes of Health, then set up shop in Bethesda and floated a massive initial public offering.

Instead of reviling his creation he would have plastered it on fliers touting his innovative technical breakthrough and announcing the launch of a new longevity clinic. The monster, meanwhile, forsaking a murder rampage as insufficiently destructive, would instead write a best-selling trash novel, appear on Oprah, and run for governor of California. Victor himself would not seek death on the ice floes of the Arctic but would look forward to a morally and financially untroubled retirement in Palm Springs as soon as his lawyers got rid of those meddlesome do-gooders from the Securities and Exchange Commission.

The fashion for tolerating ignorance of important scientific things is motivated not only by economics but also by politics. Scarcity of knowledge is considered good in some circles because it prevents wicked scientists from doing mischievous things like making babies with three heads or developing diseases that could wipe us all out in a few weeks. Whether it actually does so is debatable, of course. Laboratories around the world are now routinely cloning monkeys and farm animals, and, one presumes, people in secret. Deadly organisms are made all the time by governments for military purposes. The ease of doing so is illustrated by the celebrated case of the lethal mousepox variant created accidentally by Ron Jackson and Ian Ramshaw in 2001.[7] The potential danger of actually understanding life is increasingly cited as a justification for enacting strong new laws regulating dissemination of biological information.

Watching the classification censors swoop down on life science is another strong déjà vu experience, flashing many of us back to the time the public record of nuclear physics was cleansed. The Atomic Energy Act of 1954 states that "the development, use, and control of atomic energy shall be directed so as to promote world peace, improve the general welfare, increase the standard of living, and strengthen free competition in private enterprise." What this means is that it is now a felony to reveal certain facts about nature in public, or even to mention what facts you cannot reveal. An entire body of knowledge has been whited out. The present eradication campaign is even more earnest than the one fifty years ago, however, because bioweaponry, the nuclear technology of our age, cannot be controlled through means of production. Unlike fissile fuel, which is expensive and difficult to obtain, genes can be modified for a few dollars. The sense of security generated by all this book burning, however, is almost certainly illusory. No less a person than Edward Teller argued that nuclear classification has been ineffective, in that its long-term consequence has been to keep information from people who might use it peaceably but not from the bad guys determined to acquire it through espionage.[8] This argument is consistent with what I learned anecdotally at Livermore over the years about nuclear weapons programs in other countries—including ones that do not yet have these weapons. My colleague Jay Davis, for instance, one of the weapons inspectors in Iraq, reported that access to nuclear secrets there was "not an issue."

Beneath the economic pressures on engineering and the inherent danger of knowledge is a genuinely fascinating question of cause. Absent a coordinated worldwide conspiracy to prevent understanding of gene regulation, for which I see no evidence, one is moved to ask *why* it is so difficult to understand. The conversion of genes to protein has two steps, transcription of DNA into messenger RNA and the subsequent translation of this RNA into protein. The latter is perfectly deterministic and simple, for it boils down to a handful of control commands issued by the RNA to the ribosome, the little machine that generates the protein. Large numbers of experiments have demonstrated that the ribosome reads its instructions mindlessly and does what it is told. But nature has seen fit to make the transcription instructions so much more flexible and abstruse that experts cannot even agree on what they are. Why nature did so is not known, but the reason must be fairly important because there is no known exception in any organism. The sheer size of the global microarray experimentation budget—about one billion dollars per year—indicates just how intransigent the problem is.[9]

Philip Anderson draws a wonderfully acerbic analogy between situations of this kind and a murder mystery in which a hapless detective sleuths away while people are dropping like flies all around him. The detective is so obsessed with minutiae that he cannot see the biggest clue of all, the steadily growing pile of dead bodies in the center of the room, despite constantly tripping over them. The big clue in this case—Sherlock Holmes's dog that did not bark in the night—is the intractability itself.[10] A perfectly obvious explanation for this intractability, particularly its universality, is that biological regulation, one instance of which is transcription, exploits the physical principle of collective instability and is thus *inherently* the domain of the Dark Corollaries. This idea is not original with me: it is implicit in a number of recent books on self-organized criticality in biology, notably those of Stuart Kauffman. However, my version has a slightly different spin, in that it identifies the experimental confusion itself as a key effect, and implies that a purely deductive microscopic understanding of gene regulation may be fundamentally impossible, at least with present-day experimental strategies. Collective instability would create a Barrier of Relevance capable of destroying the predictive power and falsifiability of theories, and it would also fool people, through the Deceitful Turkey effect, into thinking they had found explanations for things when they actually had not. In other words, the machinery of life is rendered inaccessible by the very physical principles central to its function. This being the case, nature itself is the censor, not legislators or bureaucrats.

The relevance of collective instability to regulatory control is somewhat counterintuitive, so let me explain it in more detail. Consider an airplane autopilot.[11] Even though most planes are designed to fly stably, they can still be perturbed by small gusts of wind and blown off course. The autopilot is a robot that uses feedback to correct these errors. On-board gyros detect the reorientation of the plane and generate small electrical signals in response. These small signals then feed into amplifiers, which then turn them into large signals, which then actuate the control surfaces and correct the error. The amplifier is the crucial component, for the tiny physical forces sensed by the motion detectors are not remotely powerful enough to thrust big clunky control surfaces into the airstream.[12] But this amplifier is the unstable thing. Turning small signals into large ones is no different physically from responding violently to small stimuli. The amplifier's capriciousness is normally held in check by the designer, whose job it is to make the autopilot work properly, but a few misplaced wires or faulty connections can make it vibrate out of control or slam the rudder full over and crash the plane. These effects are the mechanical analogues of cancer, a disease in which a handful of tiny genetic defects in one cell amplify through the regulatory machinery of your body and kill you. Amplifiers in autopilots are made of transistors, solenoids, hydraulic valves, and so forth, but that is only because these particular components are cheap and easy to use. Any other unstable physical system would serve just as well. One could imagine, in particular, using the stiff competition between two or more organizational states, such as two kinds of crystalline order, two kinds of magnetism, or two kinds chemical reaction organizations, to make the violent sensitivity to perturbations characteristic of amplification. Collective instability, in other words, is nature's amplifier. From a functional perspective there is no difference at all between naturally occurring collective instability and the behavior of an inexpensive amplifier chip you can buy in an electronics store.

Amplification instability is an especially pernicious and effective sower of scientific confusion because it tends to disappear, like a mirage, when experiments become crude. When the autopilot is functioning, so that the airplane is locked on course, the plane's behavior reveals no trace of the underlying amplification instability. Only when one attempts to take the plane apart to see how it works does one discover the amplifier. In this respect it is like my favorite physics problem, the high-temperature superconductor. The behavior of the superconductor, like that of the plane, is perfectly understandable until one takes it apart to find a Pandora's box of complexity and confusion, at least some of which is caused by proximity to a nearby phase transition and the concomitant Dark Corollaries.

Whether such corollaries are at work in living things is not known, but the mere suggestion that they are has extremely disturbing implications for experimental biology. It places the burden of proof on the scientist to show that his or her experiment has meaning—something not commonly done at present, and even considered slightly disreputable—since measuring first and asking questions later has the potential to generate massive amounts of information that is not even wrong. It impugns the common practice of not repeating and checking experiments, since variability need no longer be natural but a symptom of instability. It devalues truth determined by consensus to the status of politics and raises the possibility that the consensus is simply enshrined and legitimized falsehood. It transforms proprietary secrecy into a golden opportunity for fraud.

Most important of all, however, the presence of such corollaries raises the concern that much of present-day biological knowledge is ideological. A key symptom of ideological thinking is the explanation that has no implications and cannot be tested. I call such logical dead ends antitheories because they have exactly the opposite effect of real theories: they stop thinking rather than stimulate it. Evolution by natural selection, for instance, which Charles Darwin originally conceived as a great theory, has lately come to function more as an antitheory, called upon to cover up embarrassing experimental shortcomings and legitimize findings that are at best questionable and at worst not even wrong. Your protein defies the laws of mass action? Evolution did it! Your complicated mess of chemical reactions turns into a chicken? Evolution! The human

Your mess of chemical reactions turns into a
chicken.

brain works on logical principles no computer can emulate? Evolution is the cause! Sometimes one hears it argued that the issue is
moot because biochemistry is a fact-based discipline for which theories are neither helpful nor wanted. The argument is false, for theories are needed for formulating experiments. Biology has plenty of
theories. They are just not discussed—or scrutinized—in public.
The ostensibly noble repudiation of theoretical prejudice is, in fact,

a cleverly disguised antitheory, whose actual function is to evade the requirement for logical consistency as a means of eliminating falsehood. We often ask ourselves nowadays whether evolution is an engineer or a magician—a discoverer and exploiter of preexisting physical principles or a worker of miracles—but we shouldn't. The former is theory, the latter antitheory.

Since collective instability is emergent, it is reasonable to ask at what scale collective principles of organization begin to matter in life. This question turns out to be impossible to answer crisply because emergence at intermediate scales is inherently ill-defined. Macroscopic emergence is identified as something universal, like rigidity, that becomes increasingly exact in the limit of large sample size, hence the idea of emerging. There is nothing preventing organizational phenomena from developing at small scales, but it is not generally possible to *prove* their existence because they are not yet exact.

There is considerable circumstantial evidence that both stable and unstable emergence occur already at the scale of individual proteins. Very large textbooks have been written on this subject, and I must refer committed readers to one of these for a comprehensive discussion.[13] The simple fact that proteins are big, for example, suggests that to work effectively they need to exhibit something analogous to mechanical rigidity, an emergent property that occurs only in systems that are large. A specific example of ideas about rigidity applied successfully to protein behavior is the functioning of ATP synthase, a little electric motor with a rotor and stator embedded in the wall of a mitochondrion.[14] Largeness is also implicated in the integrity of DNA transcription and replication, which for unknown reasons defy the usual requirement of statistical branching in chemical reactions. The idea of unstable amplification is implicated in the conversion of ATP to mechanical energy of linkages in motor proteins, such as the actin-myosin complex of muscle or kinesin,[15] as well as in the functioning of ion channel proteins and cell-surface receptors.[16]

Unfortunately, this evidence is insufficient to clear up the controversy one way or the other, which accounts for the strange effect one often finds at a genomics or protomics meeting in which the speaker switches smoothly from reductionist ideas to collective ones as convenient, just as one might switch from cards to competitive psychology when playing poker. Thus the presenter of a paper reports writing a computer program based on fictitious laws of motion for the atoms, and then using this program to predict the shapes of proteins from the underlying DNA sequence. That this strategy works at all (which it does some of the time) indicates that the particular protein's folded-up structure does not depend sensitively on the details of the interatomic forces, since if it did, one would have to implement a *correct* solution of the *correct* equations of motion. Yet if one asked these same people, or their grant monitors, whether they believed universal principles were at work, so that one could speak sensibly of "hemoglobinness" or "ribosomeness," most of them would say no.

Insofar as collective behavior does occur at the protein scale, its main importance is to add weight to the argument that collective principles are at work where it really counts—the level of systems and large-scale processes, such as metabolism, gene expression, and cell signaling, all of which are difficult to measure directly. That, in turn, requires us to take the Dark Corollaries seriously, and in particular to face the possibility that it may be fundamentally impossible to figure out the principles of life with bad experiments, no matter how much money one throws at them or how much data they generate.

Unfortunately, we must also live with the world's formidable determination to solve problems the wrong way. I was drafted into the Army by President Nixon in 1972—out of Berkeley, which makes the story even better. After basic training at Fort Ord, I was ordered to missile school in Oklahoma, a welcome turn of events, since the Vietnam War was still killing people at a rapid pace. Everything that year seemed to be running madly backward, and my trip east was no exception. My father drove me down to Los Angeles, gabbing insufferably the whole way about how it would all be over before I knew it and how obeying the law was a good thing, which only deepened my depression, and put me on a big plane to Dallas. I arrived there in the wee hours of the morning when no one was around. After searching all over the dark terminal for a cup of coffee, I eventually found one in a small shop. It was deserted except for a pair of female NCOs chain-smoking and drawling away about whupping people into shape. It was a sign. When the sun eventually rose I boarded a small prop plane, along with three other students destined for missile school, for the short hop up to Lawton. We were met there by an amiable fellow with a much thicker drawl—a refugee from Texas, apparently, since real Oklahomans do not talk that way—and led to an ancient, rusty stretch limousine with four doors on each side and a cavernous trunk. Evidently, they were expecting more people. "We'll get you there," he assured us, as though there were a chance we would vanish into the twilight zone instead. As it turned out, this is very nearly what happened. On the way through town, amid the tawdry businesses that often surround military bases, a hose under the hood broke and began flooding the windshield with water. The driver had to think fast, and his solution—I am absolutely not making this up, as Dave Barry would say—was to mumble, "We'll get you there," and turn on the wipers! We continued zooming down the road, spewing water into the air and watching the wipers fight a losing battle with the hose until, at last, it ran ominously dry. "We'll get you there," the driver said again, as the limousine began to jerk in that gluey sort of way that indicates the engine is melting. The jerks slowly escalated into lurches and then violent spasms, as he turned onto the base and headed for the personnel building, and finally culminated in one last great heave of death as the limousine pulled up to its final resting place and stopped forever. "We got you there," he said.

The pig-headed response of the science establishment to the emergent principles potentially present in life is, of course, a glaring symptom of its addiction to reductionist beliefs—happily abetted by the pharmaceutical industry, which greatly appreciates having minutiae relevant to its business worked out at taxpayer expense. The rejection of emergence is justified as defending science from mysticism. The ostensible scientific view is that life is chemical reactions, and that the bold, manful thing to do is identify and manipulate them with stupendous amounts of money and supercomputers. The corresponding mystical view is that life is a beautifully unknowable thing that can only be screwed up by humans with all their money and computer cycles. Between these extremes we have the profoundly important, but poorly understood, idea that the unknowability of living things may actually be a physical phenomenon. This does not make life any less wonderful, but simply identifies how its inaccessibility could be fully compatible with reductionist law. Unknowability is something we see all the time in the inanimate world, and it is actually not mysterious at all. Other, more primitive, systems exhibiting it have evaded computer solution up until now, and some of us are confident that they always will. Whether similar effects occur in biology remains to be seen. What is certainly true, however, is that arrogantly dismissing the possibility will lead to an endless and unimaginably expensive quagmire of bad experiments.

There is, of course, the question whether one *ought* to understand the principles of life or just pass laws requiring everyone to remain ignorant of them. Rather than just state an opinion about this obviously touchy issue, I would like to defer to one of my favorite books, Wallace Stegner's biography of John Wesley Powell, the one-armed Civil War veteran who white-watered down the Colorado River with a small team and mapped the Grand Canyon.[17] While Powell owes his place in high school history texts to this boat trip, his truly great accomplishment was the invention of government science. Powell had a great interest in the Western lands and understood that the homesteading policy tailored for weather in the East would not work in the West because of the severe multiyear droughts endemic to the region. He realized that water rights in the West were more important than land rights, and that farmers who did not have them would inevitably be wiped out. His solution was to induce Congress to authorize the United States Geological Survey, of which he was then director, to perform an irrigation survey, the hidden function of which was to modify homesteading policy on Western lands. The key moment came when he tried to evict some squatters near Clear Lake, California. The senators and congressmen from the Western states went ballistic, charging on the grounds of states' rights that Washington had overstepped its authority. Congress responded by massively cutting Powell's budget, eventually, in 1895, driving him out of office. Forty years went by with no drought. Then the Dust Bowl appeared, and all of Powell's dire predictions came to pass, including the destruction of agriculture in Oklahoma and the Depression diaspora chronicled in *The Grapes of Wrath*. Of the many morals to this story, the one relevant to scientific book burning is this: passing laws stating that physical things do not exist, when they actually do, does not work in the end. Decades of blissful happiness might ensue, but the moment of truth will come at last, and the result may well be calamitous. The right way to handle scary, dangerous things is to understand them thoroughly and deal with them openly.

As to the supposed immorality of looking at life in a mechanistic way, well, I guess I just see that as a wrong idea brought about by an excessively mechanical understanding of "mechanical." Physical law is a wonderful and astonishing creation—and vastly more impressive than its famous competitor, the human brain. The greatest disrespect I can imagine showing its Maker is to pretend that it is less competent than it is, or that it does not exist. Moreover, I happen to like machines and am rather pleased to be in their company. I prefer to be classified with them than with a lot of people I know. They are admittedly more primitive than those people, but it is wrong to hold that against them.

This talk about oneness with machines is bringing back a memory. The sun is setting at Dulles, and I am aboard a half-empty plane, sitting alone by the dark window in the back near the lavatories. The day's work is done, and I am intent on exploiting the time-zone difference between east and west coasts to sleep in my own bed. The lights in the terminal have snapped on, and trucks are scurrying about on the tarmac in the dark. The plane pushes back, then bumps and rattles across the taxiways in a sullen sort of way, a forgotten flight in a forgotten world of discarded USA Today parts and Burger Kings—the haunt of exhausted economic soldiers washing downstream like spent salmon for the bears. At the end of the runway the plane pauses, as it always does, for there is no urgency, especially when one is working late. Then, suddenly, as if called, it remembers something it forgot, its great heart begins to beat, wells of energy that are its nature and birthright are summoned, and its magnificent body is impelled forward effortlessly as it exuberantly rotates into the sky. The lights of the city recede and vanish, and the back of the plane is again dark.

I am in your hands, my young friend, and I shall trust you tonight, as I have many times before, to take me safely home.

Star Warriors

The first duty of a revolutionary is to get away with it.

Abbie Hoffman

ANCIENT GREEK MYTHOLOGY CONTAINS DEEP INSIGHTS about the human condition that make it endlessly wonderful to read and think about. For the story of the Dawn of the Golden Age of Knowledge, I refer to my well-thumbed copy of Hesiod to guard against slips of memory. It says here that fire was brought to mankind by the Titan trickster Elvis, who stole it from the gods' secret hiding place in deepest Africa. Zeus, enraged, sent a monstrous evil in the form of thousands of shy maidens, the groupies, to Elvis's brother Liberace, who was not that interested and referred them to Elvis. Zeus had given each groupie a small box full of misfortune, misery, and despair cleverly disguised as a Japanese lunch. Sure enough, each groupie eventually succumbed to curiosity or hunger and opened her box, whereupon out flew all the ills of life: phone solicitations, rush-hour traffic, televisions in airports that will not turn off, and many others. All that remained at the bottom of each box was a small jewel, a jolly personal greeting from Zeus, and the emergency hotline phone number for the Betty Ford Clinic. Not satisfied with this revenge, Zeus chained Elvis to a peeing-cherub fountain in Las Vegas and sent an eagle every day with a delivery of cocaine, barbiturates, and alcohol to gnaw at his liver. Elvis was eventually rescued from this torment by Hercules, who obtained in return the location of Zeus's marvelous golden apples and the recipe for the world-famous peanut butter and banana sandwiches that had made Elvis so big and strong. Elvis then retired to Hades, where he became immortal and fell in with some alien agents from outer space, who were there too. The aliens revitalized Elvis's career, and he now makes frequent guest appearances at abductions.

I feel guilty telling this story because, like most burlesques, it is not actually funny. Elvis Presley was a genuinely tragic hero, a person in whom the flame of creativity burned brightly, who illuminated his fellow citizens in a way they had not known before, and who died young as a result. There are countless examples of this effect from the music scene—Charlie Parker, Jimi Hendrix, Sid Vicious, Tupac Shakur—but the important point is that it is an archetype as old as mankind itself and not confined to musicians with questionable personal habits.[1]

Like Bugs Bunny, Spike Jones, and the Marx Brothers, all real theoretical physicists are anarchists. It took me a very long time to appreciate this, for I am a highly conservative person with a stable family who pays income tax and makes mortgage payments. I also studied so assiduously as a student that I had no time for politics or distractions—quite a feat at Berkeley in the early 1970s. However, the studiousness was misleading, for what I was actually doing all that time in the bowels of the library was not my homework but something funding agents in Washington viscerally hate and have come to disparagingly refer to as "curiosity-driven research"—rapid, off-line investigations of things I judged to be important. The abstruseness and abstraction of theoretical physics permit one to get away with this behavior while looking responsible, which is why the discipline is such a magnet for the independent-minded. But I failed to make the connection with anarchy until it was pointed out to me by Paul Ginsparg, the creator of the Los Alamos bulletin board, the first truly successful electronic journal of science.[2]

We were talking about why similar institutions were so slow to emerge in other branches of science. Paul suggested that physicists are self-selected to value eccentricity and novelty of ideas above all else, even at considerable professional risk to themselves. This attitude is harder to find in, say, the life sciences, which have a powerful tradition of the authority of consensus, presumably because of the great danger of someone saying something irresponsible with health implications and creating a panic. Paul felt that this cultural difference was fundamental, and that free institutions such as his would be difficult or impossible to create in any other discipline. His theory is now being put to the test, for attempts are underway to create new, electronic communication media for medicine. We will soon see whether they are genuinely new, like Paul's, or just sped-up versions of conventional journals.[3] Still, there is no denying that physicists are culturally the exact opposite of doctors.

A number of years ago I became aware of a strange effect in which extremely bright students—usually young men, but not always—would drop out of high school or college to become computer programmers. This effect is different from dropping out to do drugs, for the work is profitable, mathematically sophisticated, and beyond the abilities of most people, including most high school mathematics teachers. It is nonetheless scary, especially for parents. It went on even when I was a kid, but I had recently become sensitized to it on account of having sons in the vulnerable age bracket. (I have been spared thus far.) It struck me as an extremely odd thing to be happening in the information age, especially since it happened so frequently. I know several cases personally and many more by anecdote. Each of the handful of individuals I actually met was personable, well adjusted, and sharp. Something had just happened to alienate them—something they did not wish to talk about.

One of these people was a housemate of mine when I was a graduate student at MIT. To put his formidable abilities in perspective, he was working at the time at Bolt, Beranek, and Newman, a defense contractor, on a little thing called Darpanet—the progenitor of the internet—and later emigrated to Silicon Valley, where he makes vastly more money than I do.[4] Another was the son of a colleague. Yet another I learned about by accident at a barbecue in Los Altos Hills— the Beverly Hills of tech. I was asking a local computer entrepreneur how he found programmers for his startup, and he said he just heard about them by word of mouth. In fact, his best guy was twenty, had no degree, and was working his heart out for the princely sum of twenty thousand dollars a year. There were also rumors, which I never verified but found very credible, that half the operating staff at the San Diego Supercomputer Center had never finished college—and were vastly more capable than the people who had.

I am increasingly convinced that these frequent instances of academic meltdown are simply anarchists who slip out of the system early on account of being pushed too far in the competition for "excellence." In other words, the effect is a brainy cousin of drug abuse or teenage suicide. I probably escaped only because I grew up in a small country town where the competition was less severe.

It is a famous old bromide that financial security and professional staying power require single-minded focus on markets, competition, and conformity. All good parents understand this fact, and only the most irresponsible of them counsel or countenance anything else. I am no exception, as my sons will sadly attest. But the truth is that sometimes the imperative to focus fails, and there is absolutely nothing anybody can do about it. The impulse to live in creative freedom is powerful within all of us, and a handful of people always wind up succumbing to it despite the warnings. Whether this impulse is cultural or genetic could be debated for weeks, but what is certain is that it is the true source of art, scientific discovery that counts, and the powerful drive to innovate characteristic of modern civilization. Parents watching a child go his own way send up prayers begging whoever might be listening to make him safe. I send up a prayer too: Dear Heavenly Father, please send this guy to me.

It is probably just as well that this prayer is not answered too often. The life of an anarchist is difficult and certainly not to be encouraged. Everyone eventually grows up and has to make compromises, and indulging in willful disobedience at a young age just makes it all that much more difficult later on. For better or worse, my university's practice of admitting only very well rounded students is devastatingly effective at keeping out the rebels. One or two occasionally slip through, however, and then we can work together on some problem of importance, if only for a brief while.

The practicalities of responsible adulthood are arguably the reason discoveries tend to be made by the young. It is not that young people are smarter, although they often are, but that they have fewer promises to keep. The essence of the matter was captured by a short piece in a *Mad Magazine* drawing of a bearded hippie with flies buzzing all about, accompanied by an artfully rhymed William Gaines version of John Greenleaf Whittier: Barefoot boy with cheek of tan, no one likes a barefoot man.

Not surprisingly, many amusing things happen to anarchists when they grow up, giving rise to anecdotes that are either hilarious or bitterly cynical. For example, there is my colleague who used to argue passionately for the holy obligation of taxpayers to support groundbreaking technological research—until his wife launched a technology company and started paying taxes. He likes to salt up this story by relating how he took his daughter to breakfast at the International House of Pancakes in Sunnyvale—the heart of Silicon Valley—and overheard a theft of industrial secrets in the next booth: How much for that metal overlay procedure? Ten thousand would be fine. Here you go. How much for that diffusion process? And so forth. It took no great mathematical gifts to see that stealing this technology was vastly more cost-effective than inventing it oneself. Then there is the colleague who loved to bash lawyers in public—until it came to light that he was secretly attending law school. Another colleague performed expensive ab initio computer calculations of corrosion because his grant monitor wanted it—even though he knew perfectly well that rust was catalyzed by environmental impurities such as carbon and salt and thus could not be calculated. Yet another colleague with his career on the ropes repaired the situation by writing a sophisticated, ludicrously wrong mathematical physics paper "explaining" cold fusion and releasing it to the press, banking on the confusion in the field to cover up the deception, which it did. As a stringent test of a person's mettle, growing up easily beats calculus finals or bar exams.

The cold fusion example is dear to my heart because I was in an office with a nuclear expert when a journalist phoned him and asked for comments on the paper. It was probably the closest I have ever come to dying of a heart attack, for we were both suffocating with laughter reading the pages, each funnier than the last, as they slowly crept out of the FAX machine. But like the Elvis story, this event was actually not funny.

From a sober engineering perspective there is nothing mysterious about fusion.[5] Its allure, and thus its sales potential, comes from its being the power source of the sun and a possible source of boundless clean energy that might one day free us from dependence on all those unstable Middle Eastern countries. But it is fundamentally just an upscale version of fire—a reaction in which hydrogen nuclei combine to make a helium nucleus, analogous to the reaction of oxygen and carbon to make heat and carbon dioxide. Fusion normally takes place in violently hot environments, such as the interior of the sun, rather than cold ones, because the hydrogen nuclei repel each other strongly and thus require high impact velocities to get close enough together to fuse. Actually achieving ignition—a runaway chain reaction—is technically difficult to do without blowing oneself to kingdom come, because such high temperatures are required and such large amounts of energy are released when the reaction takes off. But it is not impossible and is therefore a legitimate technical objective of modern engineering research.

In 1989, the chemists Stanley Pons and Martin Fleischmann announced in a press conference the discovery of extra heat released in an electrochemical cell—heat they believed could be accounted for only by what they termed "cold fusion."[6] This claim made no sense at all quantum-mechanically. The energy scales of ordinary chemistry are not right for catalyzing nuclear reactions. But it turned out that enough people did not believe in quantum mechanics, were willing to distort its complexities to their own ends, or simply viewed its practitioners as con artists that the voices of reason went unheard. The Utah state legislature allocated five million dollars for cold fusion research, and there was a flurry of similar activity all over the world that author John Huizenga estimates wound up squandering between $50 million and $100 million of taxpayer money.[7]

Another very unfunny but important aspect of cold fusion—a trait it bizarrely shares with Elvis—is immortality. Back in 1997 I was driving to work one morning and happened to tune in to National Public Radio's *Science Friday* with Ira Flatow. The subject that day was cold fusion,[8] and Ira's guests were T. Kenneth Fowler, a respected nuclear engineer from Berkeley, and Eugene Mallove, editor-in-chief of the magazine *Infinite Energy*.[9] In the first half of the program, Ira elicited from Fowler a series of scholarly and scientifically responsible statements such as that one cannot tell for *sure* that cold fusion is wrong but that it is not consistent with the laws of nuclear physics as we know them and is not supported strongly by any experimental findings. The word *fraud* was avoided with great effort. But in the second half, Mallove was let loose, and his performance was impressive. I must paraphrase here to conserve space, and I apologize in advance that some of the finer shades of meaning have been lost in the condensation. Mallove said that cold fusion works, that there was experimental evidence all over the place that it did, that private capital was already using it to make power for the people, that academics had failed to make fusion profitable but were nonetheless enjoying endless bounty at the public trough, that they were the equivalent of welfare queens, and that their dumping on cold fusion was simply a campaign to squash competition and protect their jobs. He went on for thirty minutes, and Ira's failure to stop him or allow rebuttal gave me the impression that he tacitly approved.

Such episodes reveal that high-value science is often not scientific. In circumstances involving large amounts of money, being correct often matters much less than being persuasive and having sound business sense. This is one of the reasons it is so difficult to make a living as a professional theorist. Not only is being right an exercise in futility in such situations, it can get one tarred and feathered. The underlying cause is, of course, economic. One of the prices paid by people who devote their lives to fundamental discovery is to occupy a blue-collar socioeconomic niche. There is nothing wrong with blue-collar work, but it means that one is excluded from important policy meetings in which those who understand money decide what is and is not "true," often intentionally disregarding inconvenient facts coming in from the shop floor. Such annoyances go with the territory, of course, and one learns to shrug them off. But it is clear to most professional scientists that the unscientific nature of high-value science is a management effect, something embarrassing and some-times immoral but also fundamental to our lives.

Its status as a managed activity causes real science to involve signifi-cant economic sacrifice. In physics we have endured a great deal of sac-rifice recently as nuclear weapons have faded into the past, electronic hardware has migrated to the Far East, software has migrated to India, and research portfolios have shifted toward pharmaceuticals and med-icine. These shifts have been accompanied by painful readjustment—a euphemism for unemployment—for many people, and professional lives that can be nasty, brutish, and short. But there is a core cadre of

people who steadfastly endure these indignities because they understand that fundamental discovery is both possible and important—
and something that cannot be managed. The idea that it can is an
antitheory, as are the ideas that there are no more discoveries left to be
made or that the economy will magically provide them. The key break-
throughs in science have always been made by people of integrity who
went their own way, defied authority, and paid a stiff price for doing so.

   The importance of personal strength in professional life applies
not just to scientists, of course, but to everyone. On a flight from
New York to San Francisco I struck up a conversation with a fellow in
the adjacent seat, and found that he had an interesting story to tell.
He was originally from Lithuania, presently lived in Pennsylvania,
and was on his way to Alaska to go salmon fishing with his son.
When he first came to America, he said, he had gotten a job in an
electric motor factory. I am quite interested in electric motors, so I
asked him a few technical questions and discovered, to my amazement, that he knew *everything* about electric motors: the torque
characteristics, how the rotors were wound, what would and what
would not scale, the power consumption characteristics, what kind
of wire to use, heat flow from the bearings—you name it. After a fascinating hour I was sold, and joked to myself that I should become
this man's graduate student so that I could learn all about electric
motors. But a few years ago, he said, news came down from headquarters that the factory would be closed and moved to Chile. His
job evaporated. In horror I asked him what he did then. He said he
got a job in a steel mill. When I asked if that meant pushing papers in
the front office he said, no, it meant literally making steel. I am also
quite interested in steel, as it happens, so I asked him a few technical
questions and again discovered, to my amazement, that he knew
everything about making steel: what the temperature colors mean,
what the impurities should be and how one measures them, how to
properly anneal and coldwork the ingots, how to protect your crew

from injury—you name it. After another fascinating hour I was again sold, and again resolved to become this man's graduate student, so that I could learn all about steel. But a few years ago, he said, news came down from headquarters that the factory would be closed. This case was even worse than the previous one because the company had gone belly up and taken the pension fund with it. In even greater horror I asked him what he did then. He answered that by that time he and his wife had saved enough to buy a dairy, so they did, and he had run it successfully ever since. I am also quite interested in dairying, as it happens, and I was about to ask him technical questions about it that would, no doubt, have resulted in my wanting to become his graduate student to learn all about dairying, when time ran out and the plane landed. But I thought about the conversation all the way home: three times this fellow's world had collapsed, and at least twice his obviously formidable technical expertise had been rendered obsolete by economic shift. In every case he picked himself up, wrote off the investment, moved on, and thrived. He wound up fishing in a good place in the end. I believe all of us should aspire to be like this man.

I enjoy telling this story because it is a moral counterweight to the unfortunate but timeless fact that smart, well-bred people quickly tire of sacrificing and move on to other things. The moment they do is always interesting, for it reveals what they are actually made of. Sometimes one gets a heartwarming surprise, like the Lithuanian dairyman, but other times one gets something more like a used car salesman with part interest in a chop shop. Condemning the expedient life track too harshly is naive and childish, since a significant part of real business, upon which we all depend for our livelihoods, is game-playing and deception. This principle applies generally. Tender shoots in spring that do not become thorns and foxtails later in the summer will not last. Cute little kittens that do not grow up to be highly efficient killers die out fast. Rather, I observe that there are two distinct technical life tracks, one somewhat ascetic and compatible with science and the other not. The two tracks have in common a studied disregard for rules that inhibit the imagination, which can make people mistake one track for the other. But in the case of the ascetic, the rules to be disregarded tend to be intellectual, while those the salesman-scientist breaks tend to be moral. There are visionaries in each camp, and some inhabit both. For most of us, however, the choice between them is an important and difficult part of growing up.

The contrast between the ascetic and the salesman-scientist is depicted in William Broad's 1986 book *Star Warriors*,[10] the story of the notorious nuclear-pumped X-ray laser weapon program at Livermore. This project, derisively dubbed "Star Wars" by its opponents, dealt with fascinating technical issues—using laser principles to focus the enormous energy of a nuclear explosion into a narrow beam of X-rays. What makes this book memorable, however, is not the technology but the portrayal of personalities and thought processes behind a bold, high-risk engineering project. This applies to both scientists and government officials promoting the program. I knew many of these people personally, for I was at Livermore at the time, although in a lowly modeling group far removed from the thick of things. The proliferation of rule-breaking, of both kinds, is described accurately in this book just as it happened.

An important inaccuracy, however, is the scarcity of cynicism. I remember quite well how we new hires out in the trailers constantly gossiped about the physical principles involved, whether the design objectives were feasible (most of us thought they weren't), and who was almost certainly tricking whom for money. We were not that surprised when the program plummeted from its $100 million zenith in the Reagan years to zero when its promises could not be fulfilled. We were very surprised by the finale, however: we received the happy tidings that the program had destroyed the Soviet Union. That's right. They told us we had faked out the Soviets technologically and bankrupted them.

My cynicism later rose to new heights when I learned that I have been *supported* by this program. I think I would not have participated had I known, but it's difficult to judge. I had small children at home and could not afford to be unemployed even for a second.

The exaggerated claims were deliberate, of course, since the customer—in this case the Reagan administration—wanted a space-based antimissile defense and was willing to support science as a means to this specific technological end but not otherwise. In business and government, as in most things, the customer is always right, especially in matters involving very large amounts of money.

The X-ray laser program was a case study in how significant scientific progress must sometimes be made. While the laser weapon did not work, explosively driven X-ray lasers *did* work and have now become important everyday diagnostics for laser fusion. Laser fusion itself—the creation of a hot fusion environment by focusing gigantic lasers onto a tiny pellet of fuel—did not work as originally proposed because of overoptimistic assumptions about the implosion convergence, but the research *did* create the initial investments that have now led to new, conservative designs that will work—I think.[11] Whether all the financial shenanigans will be forgiven when a laser fusion pellet finally goes off remains to be seen, but the world will certainly be stunned and, I suspect, fundamentally changed.

The ability to find and exploit indirect strategies of this kind is a key characteristic of entrepreneur types I have known. Such people also tend to be sanguine about the issue of dishonesty, feeling that they have to break eggs to make omelets. I have no idea how many eggs they break per capita, but I suspect lots. I was once chatting with a technician who was wiring up a glove box (in an institution I will not name) and happened to mention that I knew a famous local visionary and operator, a fellow named X. The technician's eyes widened, and he told me in hushed tones about the time some general had called up to announce he was making a surprise visit the next day, and how X had ordered a bunch of his subordinates to stay up all night painting cardboard boxes black in order to simulate the computers he was supposed to have built but had not.

While this story evokes headshaking and smiles of disbelief from most people, it also elicits thrills of horror in *real* scientists because they are so tormented by the moral compromises required to do their work. I do not mean the moral compromises of nuclear weaponry. Involvement with these weapons is part of the tradition by which they are kept safely under civilian control, and I thus feel that no compromise at all is involved in working with them. I mean rather the need to exaggerate to get research money. Real science, as opposed to its entrepreneurial image, has a strict taboo against lying. We need this taboo to guard against wasting scarce and valuable resources, such as one's life, on false leads. Thus when Jan Hendrick Schön was recently caught falsifying a series of extremely important semiconductor experiments at Bell Labs it was perceived universally as the blackest of betrayals and resulted in wholesale hand-wringing, soul-searching, firing, and ending of careers.[12]

Because it is an essential part of our lives, the market pull of technological needs on scientific activity creates perpetual and recurring moral conundrums and is thus an inexhaustible source of dark humor. The successor of the X-ray laser as a space-based missile defense strategy was the $50 million Brilliant Pebbles—an advanced version of Smart Rocks.[13] The basic idea of Brilliant Pebbles was to deploy 4,600 small interceptors in orbit capable of homing in on and destroying enemy warheads without being steered to their targets from the ground. I do not have enough technical or military expertise to know whether Brilliant Pebbles was good idea, but I always had a problem with the guiding and aiming electronics, which are inherently difficult to make reliable. I am certain, however, that the enormous demand for missile defense was (and still is) a powerful economic incentive for at least some of its claims to be false.

By the time Brilliant Pebbles got underway in earnest in the late 1980s, I had already moved to the physics faculty at Stanford and had my hands full with departmental duties, among them chairing the qualifying examination committee. As usually happens on this committee, it got to be the middle of summer without anyone else turning in their examination problems, and it fell upon me to nag them to get it done. Being notoriously arrogant, I decided that my time was much too valuable to waste hounding these people, so I wrote the whole exam myself. This exam later became legendary among the graduate students for being impossibly difficult and full of mistakes—a truly incompetent bit of work by any measure—but it had the salubrious effect of getting me out of this particular committee duty from then on. One of the questions, in a category called general physics, was designed to test a student's ability to apply the abstract principles he or she had learned in school to practical problems such as one might encounter in the home. The previous year's problem had been to estimate the time it took to cook a pot roast. I struggled mightily with this problem, trying to think of some everyday circumstance more elegant than pot roast, but did not succeed. The longer I thought about pot roast the funnier it got, and no matter how hard I tried, I could not get it out of my head. So, in the end, I wrote a problem called *Brilliant Pot Roast*. The premise of the problem was that the United States government had a new Strategic Defense Initiative plan to orbit thousands of pot roasts in space with little tiny rocket motors attached to them, which could then be invoked at a moment's notice to deflect the roasts downward and crash them into incoming Russian reentry vehicles. The question was, what happens to you when you get hit with a pot roast going thirty-five thousand miles an hour?

After the exam was over nobody could talk about anything except pot roast. The absurdity of the problem became an editorial comment on the absurdity of our examination process itself, which I naturally had intended, and added a bit of gaiety to an otherwise dreadful two days of brutal hazing. Nobody solved the problem, of course. It was a problem in shock waves, something we do not usually teach our students until later on in their graduate careers.[14] At those speeds everything—pot roast, rock, steel—loses its shear strength and becomes a water balloon, which then splatters apart by virtue of shocks launched through it from the impacting surfaces and ricocheting back and forth inside at fantastic rates. Some of the Russian students, of which we had an especially large crop that year, came close, presumably because of a previous familiarity with military technology, but nobody actually got the right answer. A number of students complained to me that I had written down the wrong rocket equation, which I had, and several asked me what a pot roast was.[15] I said it was a piece of meat and indicated its size with my hands. I learned later that one of them had wasted a lot of time trying to estimate its mass, and observed wistfully to a friend upon leaving the exam that he had taken integral tables but should have taken a dictionary.

I did not get away with this mean trick for long. The students got together and wrote a satire modeled after Neil Simon's *Murder by Death*, which they performed at my department's Christmas party that year.[16] Its premise was that Santa Claus had been kidnapped, children everywhere were in a state of panic, and the world's greatest detectives had assembled to solve the crime. Artfully costumed, they thesped around the room bickering about who was the greatest detective of all time and who had the best theory of the crime—and finally descended on my table and accused me of mugging Santa to obtain his sleigh as the perfect launching device for my brilliant pot roast antimissile system. I was prepared. I reached under the table for the bloody roast I had hidden there and pulled it out calmly. "Ok, none of you move," I said. "This thing is armed, and I'm not afraid to use it."

The human condition is cursed by limitations and weaknesses of the mind, the sad consequences of which are everywhere around us and too numerous to name, but it also is blessed with an irrepressible instinct for optimism that wells out of people at random moments, especially when they are young. There will always be scientists—real ones—for the simple reason that there will always be a steady trickle of anarchists generated by responsible and good families doing their level best to avoid this outcome and produce only bankers, doctors, and soccer coaches. As the older ones are killed off by the practicalities of life, newer ones rise up to take their place like new grass in spring, in a cycle of creative rebirth that transcends the generations and is older than history.

There is a wonderful story from Ray Bradbury's *Martian Chronicles* that I think about from time to time because it so nicely captures this innate drive to discovery.[17] It is set in a fictional future in which Mars is beginning to be colonized, and all the colonists are curious about the ancient, dead civilization that built the nearby towns, now fallen into ruin, and where the Martians all went. One morning, a father announces quietly to his wife and two children that he has found some Martians at last, and that they would be going that day to see them. He brings the car around, and they get in and drive far out in the desert to a Martian ghost town. In the eerie silence, broken only by echoes of their own footfalls, the father leads them over to an ancient fountain and tells them to look in. The Martians are in there, he says. They look down and see no aliens at all but only reflections of themselves.

Stare into any fountain today and you will see not a Martian gazing back at you but an ancient, long dead Star Warrior, a ghostly echo of yourself living simultaneously in the future and the past. And as in Bradbury's story, you can then return home secure in the knowledge that they did indeed build those great cities, and that you have seen them.

FIFTEEN

Picnic Table
in the Sun

A human being should be able to change a diaper, plan an inva-
sion, butcher a hog, conn a ship, design a building, write a sonnet,
balance accounts, build a wall, set a bone, comfort the dying, take
orders, give orders, cooperate, act alone, solve equations, analyze a
new problem, pitch manure, program a computer, cook a tasty
meal, fight efficiently, die gallantly. Specialization is for insects.

R. A. Heinlein

IN UNIVERSITY, AS IN ANY OTHER WALK OF LIFE, THERE
are brief moments in which the loneliness of competitive professionalism falls away, and one is transported back to one's student
days, when someone else was paying the bills, and there was time to
spend talking about timeless things and an expectation that one
should do so. These moments happen rarely and do not last, for the
clouds soon cover up the sun again and return us to our practicalities. But for the brief moment while it shines, the light is a glorious
and welcome reminder of what it was like to be young. My university turns out to be a particularly good place to experience these
moments, not because they happen there more often, but because

the effect is enhanced by the stunning beauty of the place. Like the moment itself, the campus is characteristically full of light, dappled on grassy lawns and pouring out onto broad paths lined with palm trees separating earth-colored buildings with Spanish tile roofs, shimmering in the sun. How wonderful it is to sit under the shade of live oaks at a redwood picnic table with the gentle fog blowing in from the sea in midsummer. The hills are brown and quiet, the student café is still, and a few enterprising squirrels and jays are scrounging about in the dry leaves for the few undiscovered acorns. Then Gumbrecht shows up with the wine.

I had met Sepp Gumbrecht before only through email, so I was not sure what to expect.[1] The person on the other end of those emails had immense organizational ability and the kind of cultural depth you rarely find in America, so I had imagined a tall, bespectacled politician—a grandfatherly European who knew everything about life because he'd lived a long time, suffered through the war, lectured at many famous universities, endured six divorces, and so forth. That was quite incorrect. He was actually a short, roly-poly troublemaker with a twinkle in his eye, a fellow bohemian who had deftly mastered the art of looking more respectable than he really was. The amount of wine he was carrying in that box left no doubt.

The occasion for our interchanges was Sepp's Interdisciplinary Workshop on Emergence, one of a series of cross-cultural get-togethers he had organized over the years. The topic was partly my fault, for I'd submitted it in response to a plea he'd sent out for something we could all talk about without losing each other. This was an extremely tall order. We scientists tend to think of art, history, and so forth as interesting but too complex to be professionally useful to us, while the humanists tend to think of physics, chemistry, and so forth as interesting but too simple to be professionally useful to them. But the broad idea of sophistication growing out of primitiveness turned out to be so central to university life that it resonated strongly with everyone and eventually got adopted.

After introducing ourselves and exchanging a pleasantry or two, Sepp and I went inside the building for a standup continental breakfast with the cast of characters he had invited to his workshop. Some were familiar faces, but most were not, for part of being a successful university troublemaker is knowing how to shake loose bits of travel money for visitors. Right by the entrance sipping coffee was one of the locals, the cosmologist Andrei Linde,[2] whose wicked sense of humor and love of philosophy (he is Russian) brighten up my department meetings. Next to him, bioethicist Sandra Mitchell[3] stood munching a muffin and looking around for an argument to win. Over in the corner, materials engineer John Bravman[4] was vigorously defending the inevitability of nanoelectronics and micromechanical machines to religious scholar Catherine Pickstock,[5] who smiled disarmingly while strategically diverting the conversation to existentially ecstatic modes of being. Carl Djerassi, inventor of the birth control pill, discoursed nearby in an animated way about his favorite subject, sex.[6] Philosopher Martin Seel, wearing his coat European-style over his shoulders, stood in another corner commenting thoughtfully about Heidegger.[7] Anthropologist Denise Schmandt-Besserat stood nearby taking occasional small bites from a bit of cantaloupe on a toothpick and opining in a lilting French accent about the Palestinians.[8] Behind her, lawyer Rich Ford[9] was rapidly downing orange juices and fielding ideas coming his way in short, machine-gun-like sentences designed to prevent the defense from ever regrouping. Computer guru Terry Winograd was behind him explaining artificial intelligence patiently to some computationally challenged person.[10] Italian scholar Bob Harrison was mingling about and engaging in philosophical smalltalk as a means of making stealth assessments of people. Philologist Andreas Kablitz, visibly suffering from jet lag, was standing to the side describing in great detail the similarities between his recent plane ride from Europe and Dante's *Inferno*.[11] Beside him, Dean Wlad Godzich discoursed in a friendly but authoritative way about Middle East politics.[12]

Humanists have a strange practice of organizing discussions around words rather than things—the exact opposite of how it works in physical science—presumably because their business is understanding how people work rather than understanding how machines work. Accordingly, after breakfast everybody waded in with his or her understanding of the term "emergence," in the process nicely imitating Monday morning in the pits of the Chicago Mercantile Exchange. After two hours of fine statements like "emergence is the act of emerging" it became clear that this concept was strangely slippery, like Justice Potter Stewart's famous characterization of pornography: I can't define it but I know it when I see it.[13] But as the workshop progressed, things slowly came into focus, and a few strong examples eventually distilled out.[14]

The one I remember most clearly, thanks in part to Edwin O. Wilson's proliferating television specials, is the self-organization of social insects.[15] In response to some comments I made about self-assembly of atoms, Sandra pointed out that beehives also have no executive— no individual who decides who will do what or how the total economy will be structured. The bees just organize themselves. The nature of a colony is encrypted in the behaviors of individual bees, just the way the structures of Conway's *Life* are encrypted in its simple rules of motion, and are similarly difficult to anticipate. The colony thereby acquires a meaning that transcends that of its parts, just as the structures of simple cellular automata do.

Thinking about bee economies leads naturally to thinking about human economies, so we then did that. Here I got some embarrassing remedial education from Gumbrecht and Seel, who pointed out that these ideas had anti-Marxist overtones. The fundamental premise of socialism is that known rules of human behavior, insofar as they are understood, ought to be controlled by governments for the mutual benefit of everyone. But this thinking is woefully incomplete if the economy actually runs through sophisticated principles of organization encrypted in the rules of human behavior so efficiently that you cannot infer them. In this age of proliferating McDonald's and mountains of Chinese products in Wal Mart, one often hears that economies are "too complicated" to be micromanaged. This, however, is no different from saying that certain chemical processes are "too complicated" to be microscopically controlled. Thus it boils down to an acknowledgment that the essence of an economy is not in the fundamentals—food shelter, transportation, health, and so forth—but in higher-level organization that grows out of them.

Having thus explained the world economy, we were inspired to move on to consciousness. As commonly occurs when you do this, the conversation quickly got bogged down over whether consciousness is material. Professor Pickstock argued that it wasn't, and that the opposite view was simply an ideological justification for immoral behavior. Winograd countered that this was ridiculous, that the mind had to be material, and that there was nothing immoral about understanding how it works. I was on Terry's side and so restrained myself from bringing up Duke Nuke'em, internet porn, and spam. He pointed out—correctly in my view—that the acid test of materiality of the mind is building a machine that exhibits consciousness. He also admitted that so far all attempts do so had fallen well short of the goal, and thus that the jury was still out on Professor Pickstock's assertion. He said that computer scientists currently think that the failure is technical and has its roots in the fundamental impossibility of micromanaging everything a computer does. If consciousness were contained not in the program itself but were the result of emergent self-organization of logical structures it generates, you would be able to build conscious machines only after you had fully comprehended the relevant organizational principles. He said that this thinking was the impetus for the modern movement to create programs capable of "adaptive" behavior—changing their rule base in response to consequences of previous actions.

From logical structures constituting the mind we segued into logical structures the mind makes, starting with the particularly well documented case of jurisprudence. Rich told us about a debate currently raging among legal scholars as to whether legislation is objective. English and American jurists, he said, typically subscribe to the "rationalistic" view that legal disputes can, and should, be decided logically according to overarching principles or policy objectives. But the laws legislatures actually write are often vague, a practice that gives the courts enormous latitude to determine a law's actual meaning. Only after a trail of precedent has been generated can such laws be applied with predictable results. From the rationalistic viewpoint, this de facto power of the courts to make law is a symptom of bad legislation, something you want to hunt down and eliminate. But a growing group of "irrationalist" scholars believes that far from being pathological, such vagueness and its subsequent tasking of the legal system to bring the law into focus are essential to the nature of all legislation. They argue that a completely logical development of law is neither possible nor desirable.

The law discussion had the ancillary effect of getting Carl hopelessly exercised about technical regulation, which has directly impacted his great invention, as well as many other things he cares about deeply. He fumed that regulatory law was often capricious and ill-conceived, and that the root cause was that legislators did not adequately understand the social dynamic that causes important technical innovations to have large unpredictable consequences. A specific case he cited was the increasing trend for fertile young women to have their eggs extracted for preservation and fertilization years later—and thus to postpone childbearing past the dangerous age of 38 years—which was an unforeseen consequence of a treatment for infertility. He argued that it is simply not possible to control such developments, and that a best course of action for our society is to let them play out and then tailor regulatory law to the historical situation, not the other way around.

Carl's spirited lawyer-bashing emboldened Denise to observe that a textbook example of his dynamic—and also an important organizational creation of the mind—is the invention of writing. The historical facts surrounding this subject are unfortunately controversial.[16] Some scholars contend that writing originated in Mesopotamia around 3,300 BCE and then spread to the rest of the world from there. Others believe that it developed more or less independently in at least three different locations—the Near East, China, and Mesoamerica. But Denise explained that in the ancient Near East, at least, there is credible evidence that writing exploded out of an advance in counting technology. This was an agricultural society in which the ability to count was a prerequisite for survival. At a certain point, she said, someone decided to use tokens, molded from clay into simple geometric shapes, to represent goods. This first, basic, step then led to the beginnings of writing through a snowballing series of challenges and responses.

The discussion of writing evolved into a wide-ranging interchange on where language itself comes from. That development took place much earlier than writing, of course, and is correspondingly more difficult to address factually. But Wlad pointed out that that there is a fascinating clue in the historical record. Prose is taught today as a natural way of organizing writing and thinking, but this is incorrect. It developed from poetry, rather than the other way around. Poetic rhythms and rhyme, the patterns of which made things easier to remember, arose first, and were only subsequently transformed into prose.

Jumping among these and other mighty topics all morning had the expected effect of causing us to tire, lose traction, and eventually retreat to such untaxing subjects as the meaning of life. However, even this turned out to be pertinent. Sandra pointed out that the ambition of planning a life is notoriously uncertain, since the surprises that come one's way—illness, divorce, childbirth, job loss—have large effects one cannot predict in advance. Most of us understand intuitively that good mental health requires rolling with the punches and responding to such events flexibly, building up one's life history in the process. This commonsense idea is neatly captured in the title of Martin Seel's book *Sich Bestimmen Lassen*, or let yourself be determined. Its point is that in the realm of human activity (praxis) some things cannot be controlled but must be allowed to happen on their own.

At any rate, by noon nobody's brain would hold any more, so we gratefully adjourned outside to the warm sunshine for a catered lunch, complete with the stupendous supply of wine Sepp had brought in that box. Naturally this lunch lingered, naturally we all went back for refills, and naturally Sepp's box slowly began turning the morning's chaos into something coherent. It is a wonderful thing to see a master's plan unfold just the way he intended it should.

It took about an hour to happen. Casual friendly conversation slowly coalesced into an ever more serious collective effort to synthesize a definition of "emergence" from what had been discussed. Being academics, the participants just ignored the danger of getting details right and the overall picture wrong—like the parable of the six blind men trying unsuccessfully to synthesize a meaning of "elephant" from individual measurements of the trunk, knee, tail, and so forth—and bulled right through, at last coming up with an answer: Emergence means complex organizational structure growing out of simple rules. Emergence means stable inevitability in the way certain things are. Emergence means unpredictability, in the sense of small events causing great and qualitative changes in larger ones. Emergence means the fundamental impossibility of control. Emergence is a law of nature to which humans are subservient. In other words, this technically challenged crew of humanists had identified *exactly* the abstract principles we know, through measurement, to be at work in the primitive world. How very interesting.

I was not really surprised by this outcome, for I had felt that these parallels were self-evident and just needed the right venue to come out. It was nonetheless gratifying to witness. Exactly what it all means you could debate for a long time, but the explanation I favor is the simple one that human behavior resembles nature because it is part of nature and ruled by the same laws as everything else. In other words, we resemble primitive things because we are made of them— not because we have humanized them or controlled them with our minds. The parallels between organization of a life and organization of electrons are not an accident or a delusion, but physics.

A critically important "emergent" phenomenon we did not explicitly discuss because we were all academics who already understood the purpose of universities was the generation of fresh ideas by self-assembled groups of people such as ours. My father-in-law, armed with a gin and tonic or not, likes to point out that nobody knows why children learn to read. They just do. It is likewise so that nobody knows why a person's mind continues to evolve and grow through adulthood into old age. It just does. We academics are fond of taking credit for the particularly rapid mental advancement that occurs in young adulthood, but it is quite unjustified. This posturing is just part of the devil's bargain we make in an institution such as mine to winnow bright students from ostensibly less bright ones. This aspect of the business is necessary but hateful to most of us, and becomes increasingly so with age. No parent wants his offspring to become uncompetitive, but every parent wants his offspring to experience the good things of life, one of which is the joy of understanding things for the first time and of discovering that matters you thought were vastly different are really not different at all. I am a parent too, and I know that the proper place for living and learning, as opposed to performing, is not the classroom but the stoa, or its modern equivalent, the picnic table in the sun.

The end of this story is comfortingly anticlimactic and a much needed reality check on cherished ivory tower priorities. We made a terrific afternoon trip up into the nearby Santa Cruz Mountains to Djerassi's ranch, which he has transformed into a retreat for artists. You walk about there in his impossibly expensive sculpture garden admiring his taste, the gorgeous woodland environment, and the peace of the sun setting over the distant pacific. There is a great birth control joke there too, for its name, SMIP, stands for Syntex Made It Possible. After this excursion we gathered for a fabulous dinner at a place called Manresa in Los Gatos run by a great artist of a chef who created miniature feasts for the eye in nouvelle style, all set about by bohemian candlelight and artifacts from far corners of the world. Sepp hired a stretch limo for the occasion, thus saving us the trouble of worrying about overdoing it, which was very thoughtful. After dinner we motored back to campus, exchanged addresses and said farewells, after which I biked home in the dark. When I got home I tried to explain what had happened to my wife, who was even more interested in the humanistic aspects of the workshop than I was (as is often the case for physics spouses), but I was obviously unsuccessful. She packed me off to bed, mumbling something about a bad day.

I fell asleep dreaming about the famous story of the two-timing professor who comes home at three o'clock in the morning with clothes rumpled, hair disheveled, tie undone, and so forth, and tiptoes into the bedroom. The light snaps on. "Explain yourself," his wife demands. "Well," he replies sheepishly, "I admit it. I was out drinking with the boys and, well, I lost some money gambling, and there were women." "You can't fool me," she says knowingly. "You've been doing physics."

I have concluded from my years of scholarly labor that the story of Adam and Eve in the Bible is actually wrong. It is not true that the snake told Eve to eat the apple of knowledge, that she did so, that she then offered it to Adam, and that God sentenced them both to the torment of work and death as a result. What actually happened is that Adam and Eve ate snake in a Chinese restaurant called *Knowledge* and finished up with lychees and fortune cookies. Adam opened his cookie and read, "Here are the equations of the universe. Good luck with your calculations." Eve then opened her cookie and read, "Believe nothing this man says." Thus began the world as we know it.

The Emergent Age

Constantly regard the universe as one living being, having one sub-stance and one soul; and observe how all things have reference to one perception, and perception of this one living being; and how all things act with one movement; and how all things are the cooper-ating causes of all things which exist; observe too the continuous spinning of the thread and the contexture of the web.

Marcus Aurelius

A USEFUL RULE OF THUMB FOR LIVING A HAPPY LIFE IS not overdoing it with new ages. I am old enough to remember several of them, notably the Age of Aquarius, which was actually long gone by the time astrologers say it began: 17:35 Greenwich mean time, January 23, 1997. Promising new ages are a familiar feature of modern society, largely because most of us are optimists who believe in a brighter tomorrow and are thus easy marks for unscrupulous pitches. The Age of Aquarius, for example, did not bring enlightenment, peace, and love, as we hoped, but professional anxieties and family duties spiced up with AIDS, McJobs, lessness, and biological warfare.[1] Like a new condominium or a new car, a new age begins to look suspiciously like the one it supplanted after it weathers and depreciates a bit.

The attraction of new ages is the same as the impulse to search for Ultimate Truth, which all of us do from time to time. Just now, for example, I gave in to temptation and surfed the net. In addition to the usual Christian sites I found references to Ultimate Truth and Nirvana, Ultimate Truth and Nazis in South America, Ultimate Truth and Aliens from Outer Space, Ultimate Truth and the Koran, Ultimate Truth about Cary Grant, Ultimate Truth Online Magazine, Ultimate Truth in Russian Rock and Roll Bands, Ultimate Truth X-rated videos, and Ultimate Truth of the Capitalist Utilitarian De-Spiritualized Universe. The ultimate satire of this impulse is Douglas Adams's *Hitchhiker's Guide to the Galaxy*, in which the computer Deep Thought announces it has found the answer to the Great Question of Life, the Universe, and Everything, after 7.5 million years of hard work. The answer, it says, is forty-two.[2] The assembled scientists then learn from Deep Thought that while this is the answer, the *question* is unclear, so they instruct it to design an even bigger computer, the earth, to find the question. The earth is built and then thinks about the problem for three billion years. Unfortunately, five minutes before it is ready to reveal its solution it is destroyed by Vogons.

Ultimate truth is easy to satirize because it is a concept most of us find central to living but also quite useless, as a practical matter, much of the time. A person obsessed with ultimate truth is a person asking to be relieved of money—an archetype captured ultimately by *Candide*. Its meaning is also confused. For example, it sometimes means a moral precept such as the Golden Rule, which applies when the rules of commonsense pragmatism fail and thus determines a person's ethical core. That is clearly useful, but is subject to the criticism that it is software living in one's head and thus subordinate to the ultimate truths of chemistry and physics underneath. Other times Ultimate Truth means a common occurrence with meaning, such as the availability of parking spaces only when you don't need them. At other times it means the deeper laws of nature from which everything else flows—the confusion of which with rules of living gives absurdities such as forty-two. Thus it is in our natures to orient ourselves using absolute truth but to be confused and conflicted over exactly what it is.

One of the most interesting contributions of science to thought is the discovery that an analogous conflict occurs in nature at primitive levels. That it should do so might be argued to be reasonable, but the simplicity of certain systems enables us to go further and *prove* that it does so. While supernatural intervention is always difficult to disprove categorically, we know for certain that there is no need for it at this level, and that all of these miraculous behaviors can be accounted for as spontaneous organizational phenomena that descend from underlying law. We also know that while a simple and absolute law, such as hydrodynamics, can evolve from the deeper laws underneath, it is at the same time independent of them, in that it would be the same even if the deeper laws were changed.

Thinking through these effects seriously moves one to ask which law is the more ultimate, the details from which everything flows or the transcendent, emergent law they generate. That question is semantic and thus has no absolute answer, but it is clearly a primitive version of the moral conundrum raised by the alleged subordination of the laws of living to the laws of chemistry and physics. It shows allegorically how a person could easily master one and learn nothing whatsoever about the other. The epistemological barrier is not mystical but physical.

The conflict between these two conceptions of the ultimate—the laws of the parts or the laws of the collective—is very ancient and not resolvable in a few minutes' reflection or a casual conversation. One might say it represents the tension between two poles of thought, which drives the process of understanding the world the way the tension between the tonic and dominant drives a classical sonata. At any one time in history a given pole may be stronger than the other, but its predominance is only temporary, for the essence of the plot is the conflict itself.

Much as I dislike the idea of ages, I think a good case can be made that science has now moved from an Age of Reductionism to an Age of Emergence, a time when the search for ultimate causes of things shifts from the behavior of parts to the behavior of the collective. It is difficult to identify a specific moment when this transition occurred because it was gradual and somewhat obscured by the persistence of myths, but there can be no doubt that the dominant paradigm now is organizational. This is why, for example, electrical engineering students are often no longer required to learn the laws of electricity—which are very elegant and enlightening but irrelevant to programming computers. It is why stem cells are in the news but enzymatic functionalities are confined to the fine print on boxes of soap. It is why movies about Marie Curie and Lord Rutherford are out while *Jurassic Park* and *Twister* are in. The protagonists in these newer movies are not concerned with microscopic causes but with capricious organizational phenomena—as in, "Arrrggghhh! It's coming right *for* us!"

Ironically, the very success of reductionism has helped pave the way for its eclipse. Over time, careful quantitative study of microscopic parts has revealed that at the primitive level at least, collective principles of organization are not just a quaint side show but *everything*—the true source of physical law, including perhaps the most fundamental laws we know. The precision of our measurements enables us to confidently declare the search for a single ultimate truth to have ended—but at the same time to have failed, since nature is now revealed to be an enormous tower of truths, each descending from its parent, and then transcending that parent, as the scale of measurement increases. Like Columbus or Marco Polo, we set out to explore a new country but instead discovered a new world.

The transition to the Age of Emergence brings to an end the myth of the absolute power of mathematics. This myth is still entrenched in our culture, unfortunately, a fact revealed routinely in the press and popular publications promoting the search for ultimate laws as the only scientific activity worth pursuing, notwithstanding massive and overwhelming experimental evidence that exactly the opposite is the case. We can refute the reductionist myth by demonstrating that rules are correct and then challenging very smart people to predict things with them. Their inability to do so is similar to the difficulty the Wizard of Oz has in returning Dorothy to Kansas. He can do it in principle, but there are a few pesky technical details to be worked out. One must be satisfied in the interim with empty testimonials and exhortations to pay no attention to the man behind the curtain. The real problem is that Oz is a different universe from Kansas and that getting from one to the other makes no sense. The myth of collective behavior following from the law is, as a practical matter, exactly backward. Law instead follows from collective behavior, as do things that flow from it, such as logic and mathematics. The reason our minds can anticipate and master what the physical world does is not because we are geniuses but because nature facilitates understanding by organizing itself and generating law.

An important difference between the present age and the one just past is the awareness that there are evil laws as well as good ones. Good laws, such as rigidity or quantum hydrodynamics, create mathematical predictive power through protection, the insensitivity of certain measured quantities to sample imperfections or computational errors. Were the world a happy place containing only good laws, it would indeed be true that mathematics was always predictive, and that mastering nature would always boil down to acquiring sufficiently large and powerful computers. Protection would heal all errors. But in the world we actually inhabit, dark laws abound, and they destroy predictive power by exacerbating errors and making measured quantities wildly sensitive to uncontrollable external factors. In the Age of Emergence it is essential to be on the lookout for dark laws and artfully steer clear of them, since failure to do so leads one into delusional traps. One such trap is inadvertently crossing a Barrier of Relevance, thereby generating multiple ostensibly logical paths that begin with nearly identical premises and reach wildly different conclusions. When this effect occurs it politicizes the discussion by generating alternative "explanations" for things that cannot be distinguished by experiment. Another trap is the hunt for the Deceitful Turkey, the mirage law that always manages to be just out of focus and just beyond reach, no matter how much the measurement technology is improved. Ambiguities generated by dark law also facilitate fraud, in that they allow a thing to be labeled quantitative and scientific when it is, in fact, so sensitive to the whim of the measurer that it is effectively an opinion.

The Greek pantheon came into being through a series of political compromises in which one tribe or group, prevailing over another in warfare, would exercise its authority not by wiping out the gods of the losers, which was too difficult, but by making those gods subordinate to their own.[3] The ancient Greek myths are thus allegories of actual historical events that took place in the early days of consolidation of Greek civilization. While the "experiment" in that case was war, and the "truth" it revealed was some political reality, the psychological elements for inventing mythological laws were the same as those we use today to identify physical ones. You may feel that both are pathological human behaviors, but I prefer the more physical view that politics, and human society generally, grow out of nature and are really sophisticated high-level versions of primitive physical phenomena. In other words, politics is an allegory of physics, not the reverse. Either way, however, the similarity reminds us that once science becomes political it is indistinguishable from state religion. Under a system of truth by consensus one expects false gods to be systematically enshrined in the pantheon as a matter of expedience, and the cosmogony on occasion to become fictional, just as occurred in ancient Greece, and for the same reasons.

Greek creation myths satirize many things in modern life, particularly cosmological theories. Exploding things, such as dynamite or the big bang, are unstable. Theories of explosions, including the first picoseconds of the big bang, thus cross Barriers of Relevance and are inherently unfalsifiable, notwithstanding widely cited supporting "evidence" such as isotopic abundances at the surfaces of stars and the cosmic microwave background anisotropy. One might as well claim to infer the properties of atoms from the storm damage of a hurricane. Beyond the big bang we have *really* unfalsifiable concepts of budding little baby universes with different properties that must have been created before the inflationary epoch, but which are now fundamentally undetectable due to being beyond the light horizon. Beyond even that we have the anthropic principle—the "explanation" that the universe we can see has the properties it does by virtue of our being in it. It is fun to imagine what Voltaire might have done with this material. In the movie *Contact* the Jodie Foster heroine suggests to her boyfriend that God might have been created by humans to compensate for their feelings of isolation and vulnerability in the vastness of the universe. She would have been more on target had she talked about unfalsifiable theories of the origin of the universe. The political dynamic of such theories and those of the ancient Greeks is one and the same.

The political nature of cosmological theories explains how they could so easily amalgamate with string theory, a body of mathematics with which they actually have very little in common. String theory is the study of an imaginary kind of matter built out of extended objects, strings, rather than point particles, as all known kinds of matter—including hot nuclear matter—have been shown experimentally to be. String theory is immensely fun to think about because so many of its internal relationships are unexpectedly simple and beautiful. It has no practical utility, however, other than to sustain the myth of the ultimate theory. There is no experimental evidence for the existence of strings in nature, nor does the special mathematics of string theory enable known experimental behavior to be calculated or predicted more easily. Moreover, the complex spectroscopic properties of space accessible with today's mighty accelerators are accountable in string theory only as "low-energy phenomenology"— a pejorative term for transcendent emergent properties of matter impossible to calculate from first principles. String theory is, in fact, a textbook case of a Deceitful Turkey, a beautiful set of ideas that will always remain just barely out of reach. Far from a wonderful technological hope for a greater tomorrow, it is instead the tragic consequence of an obsolete belief system—in which emergence plays no role and dark law does not exist.

The analogy with Greek religion also applies to the humbler end of the research spectrum, where warring among scientists to see whose emergent god is more powerful is an everyday reality. A case in point is ordinary semiconduction. Back when I was in grade school, it was said that the tribe of semiconductor physicists lived in peace in the Silicon Valley and worshiped crystallinity, the daughters of which, the gods of valence band and conduction band, caused transistor action and prosperity. But then they were invaded by a hostile tribe of chemists, who worshiped not the crystal but the molecule and who believed *its* offspring, the lowest unoccupied molecular orbital and highest occupied molecular orbital, were the true cause of transistor action, and that the worshipers of the old gods were inferior and unclean. The two tribes engaged in bloody combat—fought with disinformation, dirty tricks, and refusal to speak the name of the other tribe's gods—each hoping to starve the other tribe of research dollars and thus annihilate it. The war resulted in stalemate, the vestiges of which persist today. As often happens in conflicts of this kind, the war was not really over conceptual matters at all but money, for these warring gods are actually different names for the same thing. Similar wars occur routinely in biology, although they are vastly nastier on account of the greater resources involved.

The transition to the Age of Emergence is also characterized by the increased menace of antitheories, bodies of thought that stop inquiry and thus impede discovery. Antitheories are a greater threat now because they are cheaper to generate and more expensive to destroy than they were in the past, partly because of increased demand. A world populated by proliferating laws, some of which are angels and others devils, is considerably less appealing than a world ruled by a beneficent master law, such as evolution, that makes it unnecessary to understand anything else. The master antitheory of the age is the idea that there is no fundamental thing left to discover, so that the world we inhabit is simply a swarm of detail that belongs to no one and thus can be legitimately handled by business tactics—resource management, competitive advertising, survival of the fittest, and so forth. A corollary is that there is no absolute truth, but only products, like shirts or hamburgers, that one throws away when their usefulness is exhausted. Antitheories are dangerous ideologies not only because they impede inquiry but because they lull one into ignoring threats that one's opponents can exploit to their advantage.

In the Age of Emergence, ideologies run amok more easily than they did in the past. The reason is that laws of descent are subtle and thus expensive to work out correctly, and all of us have powerful economic incentives to see these laws in a light beneficial to ourselves, even if incorrect. It takes enormous self-control to sublimate these desires, especially when one's livelihood is on the line. Ordinary mortals simply cannot do it all the time. As a result, a larger portion of the accepted knowledge base of modern science is untrue than was the case in the Age of Reductionism, obligating us to look at it more skeptically than we did before and to value consensus less.

I saw Shanghai for the first time this spring. It was the venue for a small annual meeting I have with a terrific group of Japanese colleagues I secretly call the Magnificent Seven.[4] These folks are so good that conferring with them brings me up to date on everything important in my field and relieves me of much of the need to travel. We typically have this meeting in Hawaii, but we held it in China this year as a way of helping out our friends there, and also minimizing costs. The cost containment was helped in this case by the first outbreak of severe acute respiratory syndrome (SARS), which was just beginning then. It was scary, but not debilitatingly so. We wore masks at key moments. Visiting an ethnic Chinese country is inevitably a weight-gaining experience for a westerner, for China is tied with France as the food capital of the universe. It is unthinkably inhospitable in this culture to offer important foreign guests "just enough" food. There has to be a vast surplus, and it has to be the good stuff. Thus in the Golden Temple restaurant by the waterfall with pictures of dignitaries such as President Clinton on the walls there is an endless procession of dumplings, pork in oyster sauce, bok choy with shrimp, hot Hunan special spicy chicken, and so forth, all washed down with excellent local beer. Some of my party went on to a jazz club after dinner that evening, but I was beat and decided to walk with the other group back to the hotel. The bund was all lit up with klieg lights like a Hollywood movie set, and it was thick with couples strolling about enjoying the evening. This went on until about eleven, when the lights went out, as they do every night, and loudspeakers began exhorting everyone to go home. China is a place that has suffered immeasurably from ideology and is now slowly pulling away from its effects, thanks in part to large infusions of capital from Singapore, Hong Kong, and Taipei. It has an immensely long way to go, and while this is happening, the locals are ashamed of their past and do not want outsiders to see any of it—even though we outsiders understand perfectly, for we have been in such situations ourselves.

As a result, Shanghai itself is partly real and partly an impressive show with more than the usual amount of tawdriness behind the façade. But it is also a statement: I have left my ideology behind, and this is what I shall become.

Those of us who live in industrially advanced countries know that it isn't that easy, and I suspect there is trouble ahead for these folks when the harsh realities of free-market economies finally meet those of the socialist support system. But the sentiment is nonetheless both brave and right-minded. When I was in Shanghai I mentioned this idea to one of my colleagues, a thoughtful, warmhearted man who had worked in Trieste for many years at the International Center for Theoretical Physics and was now back in Beijing. He thought for a moment and then remarked how characteristically American my observation was. He intended it as a compliment, and I took it as such.

The painful echoes of ancient Greece in modern science illustrate why we cannot live with uncertainty in the Age of Emergence, at least for very long. One often hears that we must do so, since the master laws do not matter and the little subsidiary ones are too expensive to ferret out, but this argument is exactly backward. In times of increased subtlety one needs *more* highly quantitative measurements, not fewer. A measurement that cannot be done accurately, or that cannot be reproduced even if it is accurate, can never be divorced from politics and must therefore generate mythologies. The more such shades of meaning there are, the less scientific the discussion becomes. Accurate measurement in this sense *is* scientific law, and a milieu in which accurate measurement is impossible is lawless.

The need for precision, in turn, redoubles the need for that other great Greek tradition, open discussion of ideas and ruthless separation of meaningful things from meaningless ones. Precision alone does not guarantee good law. Financing practices in the Age of Emergence have the side effect of diluting content, engendering the famous joke that the *Physical Review* is now so voluminous that stacking up successive issues would generate a surface traveling faster than the speed of light—although without violating relativity because the *Physical Review* contains no information. The problem, which is not restricted to physics, occurs because large experimental laboratories cannot get the continued funding they need without defending their work from criticism, which they typically do by forming self-refereeing monopolies that define certain ideas and bodies of thought to be important, whether they actually are or not. In extreme cases, one gets a complex web of sophisticated measurements that serve no purpose other than to expand journals and fatten frequent-flier accounts. For real progress to occur it is necessary to mix in a bit of creative destruction with one's technology. One might invoke yin and yang as a metaphor for this creative synergy, but I prefer transforming their interlocking symbol into the left and right banks of the Seine. The right bank is government and measurement, the left bank is anarchy and art, and the conflict between them is Paris. One of my French colleagues put it much better. "Yes," he said with a twinkle in his eye, "I was on the right bank once."

Back in November 1998, a month after our Nobel Prize had been announced, all the newly designated laureates and their spouses were invited to a black-tie dinner at the Swedish ambassador's residence in Washington. It was a clever move on the ambassador's part, actually, since he was using us as bait to lure the Washington scene into his home. It worked very well.

One of the fellows at my table had the label "Safire," but was shorter and more reserved than I would have expected columnist William Safire to be. So I asked him about it, and he said yes he was indeed the famous columnist. The couple to our right got an immense kick out of this and explained to my wife and me, quietly, that we would probably disagree with everything this man said but would be greatly entertained by it. It turned out that Mr. Safire knew a great deal about a great many things, including, interestingly enough, physics. He had gone to school with Leon Cooper, recipient of the superconductivity theory Nobel Prize along with John Bardeen and Bob Schrieffer, and still talked to him regularly. Then came the bombshell: Leon believed physics to be dead. He thought there was nothing of importance left to do and had moved on to modeling signal processing pathways in the brain.

At this point there was a commotion in the center of the room, and an announcement made that the after-dinner entertainment would be a game in which the new laureates would step up to the mike and field questions from the audience, suitably filtered through a master of ceremonies. So, while everyone diligently wrote out questions on little slips of paper, Dan, Horst, and I excused ourselves and proceeded to the podium. When our turn to speak finally came, the questions turned out to be the usual ones, for the most part, such as what use is your work and what will you do with the money. But then Horst got a tough one: Is Einstein relevant any more? I felt sure the question had come from Mr. Safire, given the conversation we had just had, but in any event it is a question in the popular mind. Horst was quite shaken up by it and awkwardly tried to explain that he was not "that kind" of physicist and thus unqualified to answer such a question. This was exactly the response required by etiquette at a semiconductor meeting, where extreme conservatism and pretending to be uninterested in such things is part of the professional norm, but completely inappropriate for an audience such as this one. It was also disingenuous, since all of us, deep down, are "that kind" of physicist. So I asked for permission to take the mike temporarily and deliver my version of the answer. Einstein's ideas, I said, were certainly *right*, and one sees evidence for them every day, but the deeper sense of the question had been not so much whether relativity was right as whether fundamental things mattered and whether there were any more of them left to discover. I explained that I had heard this concern voiced again and again in my travels around the world and had come to recognize it as technological hubris—like the suggestion in 1900 that the patent office should be abolished because everything had already been invented. Just look around you, I said. Even this room is teeming with things we do not understand. Only people whose common sense has been impaired by too much education cannot see it. The idea that the struggle to understand the natural world has come to an end is not only wrong, it is ludicrously wrong. We are surrounded by mysterious physical miracles, and the continuing, unfinished task of science is to unravel them. There was a brief silence after I finished, followed by a rising swell of applause—a fitting dismissal of the antitheory of the death of science. I returned to my table feeling rather good about this result, a feeling enhanced by Mr. Safire's subsequent advice that I should write a book.

The applause at the ambassador's dinner was not as miraculous as it might seem, for I have given roughly the same speech all over the world and gotten the same result. The first time it happened was not in America but in Japan. I concluded at the time that it was because Japan was a Buddhist country, but this was incorrect. I repeated the experiment in Amsterdam, and the result was almost identical, right down to the number of hands raised and the specific questions asked. Holland is about as un-Buddhist as one could possibly imagine. Then I tried it in Göteborg, Montreal, and Seoul, and the response was always the same. That there should be interest in physics in many corners of the world was perhaps not so shocking. The real surprise was its uniformity from one country to the next. The world appears to possess an enormous reservoir of thoughtful people from disparate walks of life—business, medicine, government, engineering, agriculture—who love science and understand intuitively that there is much, much more yet to come.

In passing into the Age of Emergence we learn to accept common sense, leave behind the practice of trivializing the organizational wonders of nature, and accept that organization is important in and of itself—in some cases even the *most* important thing. The laws of quantum mechanics, the laws of chemistry, the laws of metabolism, and the laws of bunnies running away from foxes in the courtyards of my university all descend from each other, but the last set are the laws that count, in the end, for the bunny.

The same is true for us. Those who refuse to see reason are invited to go with me into the high country in July, where there is not such an urgent need for quantum mechanics and elementary particles. It will not be so difficult. We will rise early on a chilly morning and light my butane stove for cocoa. No bears have come in the night, fortunately, but the reason is not our clever suspension of the food but the intelligence of the bears, who know enough to go down to the big campgrounds where the people are. We sit on cold granite admiring the sizes and shapes of flashing flecks of mica, sipping excessively hot chocolate and watching the gold sunlight paint the tips of peaks and slowly descend. A small steam burbles through the chinquapin a few feet away, keeping us company, as it did throughout the night. Grey stones lie all about in shadow on granite slabs or bare earth, covered in some places with matted pine needles. Everyone else is still asleep. The cold downhill canyon wind blows for a while and then dies away in preparation for its morning reversal. The sunlight reaches down to light up the nearby trunks one by one and finally floods the ground, eliciting muffled complaints from previously sleeping people who now understand that they will roast if they remain in those bags. Complaints give way to clomping of boots, clanging of aluminum pots, and unfocused conversation about who actually won the card game, whose job it is to cook the oatmeal, and who mislaid the toilet paper. Organizational activity then magically takes place in which the frowzy slowly transform into the clean and purposeful, paraphernalia slowly self-assemble into packs, and the ground is rendered so spotless that the chipmunks and jays wonder what happened. We then set off together through the deadfall toward the summit. There is relatively little conversation, for there is more swampy mud and skunk cabbage than we would have liked, and the rock climbing beyond the edge of the forest requires concentration. As usual in the high country, the climb up is hot in the sunshine but cold in the shade, the latter provided by granite shelves punctuated here and there with pines growing right out of the rock for no reason at all. After a long, treacherous ascent we reach the lip and discover, to our surprise, that the other side is a shallow plateau in which our stream, now a trough bulging with violently purple lupine, snakes among immense boulders toward a vast meadow carpeted with pink wildflowers. Bumblebees are there happily stuffing themselves, as is a large buck, who gets spooked by our approach and bounds away. We walk through this meadow to the head of a small lake to fill up our water bottles, wolf down a couple of peanut butter sandwiches and dried apricots, and then proceed up over the second, colder summit, now on a dusty trail well worn by hoofs of many horses. It is noon, and as we are now beginning to hanker for the steak reward awaiting us below, we redouble our efforts to reach the pack station before nightfall. The miles of flat traverses through bone-dry meadow separated by ridges of tumbled boulders suddenly transform into a dizzying descent through a cleft at the base of a basalt monolith, whose walls gush with springs that appear out of nowhere as if by magic, the waters tumbling and foaming together down to the valley below. Plodding down through dark red fir forest carpeted with springy humus and ferns and across brilliantly illuminated rock terrains, we finally reach an ocean of sage, surrounded on all sides by impenetrable mountains, the most westerly of which throw shadows that tell us the day is old. We follow the watercourse, now a raging river, through a canyon scented with cedar and Jeffrey pine, climb away up onto the rocky valley wall, so intent on making it home that we barely notice the flame of sunset light up the glacier's handiwork all about, through which it is now difficult to walk. Down into the rocky chasm through which the river roars in a torrent, across the high bridge over the noisy tumult below, difficult to see on account of the failing light, stumbling in the dark in the ruts of the ancient road blasted out of the granite by gold miners, we at last reach the meadow, then the great corral with its society of tired, satisfied pack animals, then the station itself. It is pitch dark. I shall take you through the creaky screen door into the restaurant and buy you that steak. It will be the most wonderful thing you have ever eaten.

We live not at the end of discovery but at the end of Reduction-ism, a time in which the false ideology of human mastery of all things through microscopics is being swept away by events and rea-son. This is not to say that microscopic law is wrong or has no pur-pose, but only that it is rendered irrelevant in many circumstances by its children and its children's children, the higher organizational laws of the world.

NOTES

PREFACE

1. The conflict between science and the humanities is notorious. See C. P. Snow, *The Two Cultures* (Cambridge U. Press, Cambridge, 1993).

2. Aristotle, *The Complete Works of Aristotle: The Revised Oxford Edition*, J. Barnes, ed. (Princeton U. Press, Princeton, 1995).

3. Darwin's treatise is so straightforward that it is best read in the original. See C. Darwin, *The Origin of Species*, G. Suriano, ed. (Bantam, New York, 1999).

4. The Duck's Breath Mystery Theatre, known in some circles as the American Monty Python, was formed by a group of students from the University of Iowa in 1975. They later emigrated to San Francisco, became famous for their comedy routines, and began appearing regularly on National Public Radio's *Science Friday*. Recordings and memorabilia of this group may be obtained at http://www.drscience.com.

5. B. Greene, *The Elegant Universe: Superstrings, Hidden Dimensions, and the Quest for the Ultimate Theory* (Norton, New York, 1999).

6. J. Horgan, *The End of Science: Facing the Limits of Knowledge in the Twilight of the Scientific Age* (Addison-Wesley, Reading, Massachusetts, 1997).

7. I. Prigogine, *The End of Certainty: Time, Chaos, and the New Laws of Nature* (Simon and Schuster, New York, 1997).

8. P. W. Anderson, More is Different, *Science* 177, 393 (1972).

ACKNOWLEDGMENTS

1. R. B. Laughlin and D. Pines, *Proc. Natl. Acad. Sci.* 97, 28 (2000).

CHAPTER 1

1. *Ansel Adams: American Experience*, directed by Ric Burns. For further information see http://www.pbs.org/wgbh/amex/ansel.

2. J. M. Faragher, *Rereading Frederick James Turner* (Yale U. Press, New Haven, 1999).

3. The association of science with the pioneer spirit is a core idea of a famous memorandum entitled "Science, the Endless Frontier," submitted in 1945 by Vannevar Bush to President Roosevelt, which eventually resulted in the creation of the National Science

Foundation. See G. P. Zachary, *Endless Frontier: Vannevar Bush, Engineer of the American Century* (MIT Press, Cambridge, Mass., 1999); and V. Bush, *Endless Horizons* (Ayer Co. Pub., Manchester, New Hampshire, 1975).

4. S. J. Gould, *The Lying Stones of Marrakech* (Three Rivers Press, New York, 2000), p. 147ff.

CHAPTER 2

1. The tendency of men and women to navigate differently is a notorious joke among married couples. There is even evidence from functional magnetic resonance imaging that it has a physiological basis. See G. Grön et al., *Nature* **3**, 404 (2000).

2. The bimetallic strips in household thermostats are just one of many kinds of thermometer. See J. F. Schooley, *Thermometry* (CRC Press, Boca Raton, FL, 1986).

3. The force of gravity is routinely measured by geologists to an accuracy of one part in one hundred million with instruments that use conventional weights and springs. See W. Torge, *Geodesy*, 3rd edition (Walter de Gruyter, Berlin, 2001).

4. Christiaan Huygens invented the pendulum clock in 1656 using Galileo's pendulum to regulate an existing clockwork. His first clock had an accuracy of within one minute per day. He later improved this to ten seconds per day. Huygens also invented the balance wheel and spring assembly in 1675. See J. G. Yoder, *Unraveling Time: Christiaan Huygens and the Mathematization of Nature* (Cambridge U. Press, Cambridge, 2002).

5. Léon Foucault measured Earth's rotation by swinging a heavy iron ball from a 200-foot-long wire, and he received the Copley medal of the Royal Society for this accomplishment in 1855. He also did work with Fizeau on light and heat and used a revolving mirror to measure the speed of light in various media, showing this to vary inversely with refractive index. The Foucault pendulum is described in any elementary mechanics text. See A. P. French, *Newtonian Mechanics* (W. W. Norton, New York, 1971). Building an amateur Foucault pendulum is described in C. L. Stong, *Scientific American* **198**, 115 (1958). The original reference is M. L. Foucault, "Démonstration du Movement de Rotation de la Terre moyen du Pendule," *Comptes Rendus Acad. Sci.* **32**, 5 (1851). See also http://www.calacademy.org/products/pendulum.html.

6. The best known of these experiments is the interferometric measurement of Albert Michelson, later improved by Michelson and Morley, for which Michelson received the Nobel Prize in 1907. See A. A. Michelson, *Studies in Optics* (Univ. of Chicago Press, Chicago, 1962). An excellent review of later improvements to the Michelson–Morley experiment may be found in R. Shankland et al., *Rev. Mod. Phys.* **27**, 167 (1955). The original references are A. A. Michelson, *Am J. Sci.* **22**, 20 (1881) and A. A. Michelson and E. W. Morley, ibid. **34** (1887). See also E. Whittaker, *A History of the Theories of Aether and Electricity: The Classical Theories* (Nelson and Sons, London, 1951).

7. There is a voluminous literature of postmodern philosophy of science. The most widely cited work is probably J.-F. Lyotard, *The Postmodern Condition: A Report on Knowledge* (U. of Minnesota Press, Minneapolis, 1984). See also H. Lefebvre, *Introduction to Modernity: Twelve Preludes* (Verso, London, 1995); and M. Foucault, *The Order of Things: An Archaeology of the Human Sciences* (Random House, New York, 1994). There

is also a large antipostmodernist literature. See, for example, N. Koertse, *A House Built on Sand: Exposing Postmodernist Myths about Science* (Oxford U. Press, Oxford, 1998); and the account of the Sokal hoax in A. D. Sokal and J. Bricmont, *Fashionable Nonsense: Postmodern Intellectuals' Abuse of Science* (St. Martin's Press, New York, 1998).

8. Irving Langmuir's lecture "Pathological Science" was delivered at the Knolls Research Laboratory on December 18, 1953. A transcript of this lecture may be found in R. L. Park, *Voodoo Science* (Oxford U. Press, London, 2000). See also http://www.cs. princeton.edu/~ken/Langmuir/langmuir.htm.

9. P. J. Mohr and B. N. Taylor, *J. Phys. Chem. Ref. Data* **28**, 1713 (1999); *Rev. Mod. Phys.* **72**, 351 (2000); http://physics.nist.gov/constants.

10. There is a huge literature on these effects, known collectively as quantum electrodynamics. The most celebrated of them is the Lamb shift. See W. E. Lamb and R. C. Retherford, *Phys. Rev.* **79**, 549 (1950); ibid. **81**, 222 (1951).

11. A. H. Guth and A. P. Lightman, *The Inflationary Universe* (Perseus Publishing, Cambridge, Massachusetts, 1998).

12. The Josephson and von Klitzing constants are $K_J = 2e/h$ and $R_K = h/e^2$, where e is the electron charge and h is Planck's constant.

13. The literature on Deng Xiaoping is vast and, in light of historical events, complicated. See M. J. Meisner, *The Deng Xiaoping Era: An Inquiry into the Fate of Chinese Socialism, 1978–1994* (Hill and Wang, New York, 1996).

14. Murphy's law states that if anything can go wrong, it will. According to the U.S. Air Force Flight Test Center History Office, Murphy's law was born at Edwards Air Force Base in 1949. It was named after Captain Edward A. Murphy, an engineer working on a project for determining how much sudden deceleration a person can withstand in a crash. See A. Bloch, *Murphy's Law and Other Reasons Why Things Go Wrong* (Price Stern Sloan Pub., Los Angeles, 1977); and http://www.edwards.af.mil/history/ docs_html/tidbits/murphy's_law.html.

CHAPTER 3

1. Newton's treatise *Philosophiae Naturalis Principia Mathematica* is reprinted regularly. See, for example, I. Newton, trans. by I. B. Cohen and A. Witman, *The Principia: The Mathematical Principles of Natural Philosophy* (U. of California Press, Berkeley, CA, 1999). There are also numerous biographies of Newton and anthologies of his work in print. See R. S. Westfal, *The Life of Isaac Newton* (Cambridge U. Press, Cambridge, 1994); and B. I. Cohen, *Newton: Texts Backgrounds Commentaries* (W. W. Norton, New York, 1996).

2. The term *clockwork universe* has a somewhat pejorative flavor nowadays. See S. J. Goerner, *After the Clockwork Universe* (Floris, Edinburgh, 1999).

3. That comets actually travel in highly elliptical orbits and return periodically was discovered by Edmund Halley, who used Newtonian mechanics to predict the return of the comet that now bears his name. See C. Sagan and A. Druyan, *Comet* (Ballantine, New York, 1997). The original reference for Halley's discovery is E. Halley, *Phil. Trans. Royal Soc. London* **24**, 1882–1899 (1705).

4. Neptune's orbit was "predicted" by Adams and Leverrier and discovered by Galle. See S. Drake and C. T. Kowal, *Scientific American* **243**, 52 (1980); and P. Moore, *The Planet Neptune* (Wiley, Chichester, 1988). Pluto was predicted by Percival Lowell and discovered by Clyde Tombaugh in 1930. See S. A. Stern and D. J. Tholen, *Pluto and Charon* (U. of Arizona Press, Tuscon, 1998).

5. John Harrison invented the first marine chronometer, which he called H–4, in 1759. It was essentially a large watch, four inches in diameter, with a bimetallic strip for a spring. It was first tested in 1762 on a six-week voyage from England to the Caribbean. The reported error upon arrival in Jamaica was only five seconds. The corresponding longitude error of 1.25 minutes of arc, or absolute navigational error of 30 miles, easily qualified Harrison to win the longitude prize endowed by the London Board of Longitude. For complex reasons, the board paid him only a fraction of the £20,000 prize money, and King George III had to intervene personally on his behalf to cause the rest to be released. One of Harrison's first chronometers accompanied Captain Cooke on his second voyage (three years) ending in 1776. Cooke called the chronometer ". . . our faithful guide through all the vicissitudes of climates." See D. Sobel, *Longitude* (Walker and Co., New York, 1995).

6. A nice discussion of atomic clocks may be found in C. Audoin, B. Guinot, and S. Lyle, *The Measurement of Time* (Cambridge U. Press, Cambridge, 2001).

7. The first seaworthy gyrocompass was made in 1908 in Germany by the firm of Hermann Anschütz, using principles worked out by Max Schuler. Elmer Sperry invented a cheaper gyrocompass in 1911, and also invented the gyroscopic ship stabilizer. See T. P. Hughes, *Elmer Sperry: Inventor and Engineer* (Johns Hopkins U. Press, Baltimore, 1993).

8. S. Drake, *Galileo at Work: His Scientific Biography* (U. of Chicago Press, Chicago, 1978).

9. The telescope was invented in the Netherlands. The national government in the Hague discussed in October 1608 a patent application by Hans Lipperley for a low-power telescope equivalent to modern opera glasses. Within a year such telescopes were available for sale in Paris. The first astronomical telescope was made by Galileo in 1609. With it he discovered the moons of Jupiter and resolved nebular patches of stars. See H. King, *The History of the Telescope* (Griffin, London, 1955).

10. Excerpts from *The Assayer* (*Il Saggiatore*) and other writings of Galileo may be found in S. Drake, ed., *Discoveries and Opinions of Galileo* (Barnes and Noble, New York, 1989). See also S. Drake and C. D. O'Malley, *The Controversy of the Comets of 1618* (U. of Pennsylvania Press, Philadelphia, 1960).

11. There are several excellent publications on the matter of Galileo's trial and imprisonment for heresy. See P. Redondi, *Galileo Heretic* (Princeton U. Press, Princeton, 1987). A particularly thoughtful discussion may be found at the International Catholic University website: W. E. Carroll, *Galileo: Science and Religion*, http://www.catholicity.com/school/icu/c02907.htm.

12. Galileo Galilei, trans. by S. Drake, *Dialogue Concerning the Two Chief World Systems* (U. of California Press, Berkeley, 1967).

13. It is open to interpretation whether this statement is correct. At the time, Italy had converted to the modern Gregorian calendar, but England was still using the Julian calendar. Therefore, although Newton's birth certificate and Galileo's death certificate

both give the year 1642, according to the Gregorian calendar, Newton was born on January 4, 1643, while Galileo died on January 4, 1642, while according to the Julian calendar, the respective dates are December 25, 1642, and January 4, 1643. See http://home.att.net/~numericana/answer.

14. J. B. Brackenridge, *The Key to Newtonian Dynamics: The Kepler Problem and the Principia* (U. of California Press, Berkeley, 1995).

15. For an accessible (for westerners) review of Buddhism see D. C. Conath, *Buddhism for the West: Theravada, Mahayana, and Vajrayana* (McGraw-Hill, New York, 1974).

16. The best of the many books on chaos is by its discoverer: E. N. Lorentz, *The Essence of Chaos* (U. of Washington Press, Seattle, 1994). See also J. Gleick, *Chaos: Making a New Science* (Penguin, New York, 1998); and G. P. Williams, *Chaos Theory Tamed* (Joseph Henry Press, Washington, D. C., 1994).

17. This particular false syllogism comes from the home page of Golden Gate University: http://internet.ggu.edu/university_library/if/false_syllogisms.

18. For overviews on neutral helium atom diffraction from surfaces see G. Scoles, ed., *Atomic and Molecular Beam Methods, Vols. I and II* (Oxford U. Press, New York, 1992); and D. P. Woodruff and T. A. Delchar, *Modern Techniques of Surface Science* (Cambridge U. Press, New York, 1994). The original reference for the discovery of atom diffraction is I. Estemann and A. Stern, *Z. Physik* **61**, 95 (1930). See also http://sibener-group.uchicago.edu/.

19. For a comprehensive overview of electron diffraction see J. M. Cowley, ed., *Electron Diffraction Techniques* (Oxford U. Press, New York, 1992). The original reference for the discovery of electron diffraction is C. J. Davisson and L. H. Germer, *Phys. Rev.* **30**, 705 (1927). For a discussion of modern technology see A. Tonomura, J. Endo, T. Matsuda, and T. Kawasaki, *Am. J. Phys.* **57**, 117 (1989).

CHAPTER 4

1. Ice fishing is enormously popular, and much information about it is available for free on the Internet. See, for example, http://www.icefishingworld.com and http://www.invominnesota.com. See also J. Capossela, *Ice Fishing: A Complete Guide, Basic to Advanced* (Countryman Press, Woodstock, VT, 1992).

2. See http://icefishingoutdoors.com/safety.html. Mr. Smalley can be reached at tim.smalley@dnr.state.mn.us or http://www.dnr.state.mn.us.

3. For the current status of first principles calculations of the properties of water, see T. R. Truskett and K. A. Dill, *J. Chem. Phys.* **117**, 5101 (2002) and references therein. The phase diagram of water is still not completely known, even experimentally. The controversies are described in C. Lobban, J. L. Finney, and W. F. Kuhs, *Nature* **391**, 268 (1998). See also F. Franks, *Water: A Matrix of Life* (Royal Society of Chemistry, Cambridge, 2000). Useful Internet references on the phase diagram of water include http://www.sbu.ac.uk/water/phase.html and http://www.cmmp.ac.uk/people/finney/soi.html.

4. The literature on physical chemistry is so vast that there is no good comprehensive review. A good place to start is W. J. Hehre, L. Radom, P. V. Schleyer, and J. Pople, *Ab*

Initio Molecular Orbital Theory (Wiley, New York, 1986). A good introductory text is A. M. Halpern, *Experimental Physical Chemistry* (Prentice-Hall, Upper Saddle River, New Jersey, 1997).

5. The most famous demonstration of a phase transition arising from simple rules is the Onsager solution of the 2-dimensional Ising model. It is explained in detail in K. Huang, *Statistical Mechanics* (Wiley, New York, 1963), p. 349ff. The original reference is L. Onsager, *Phys. Rev.* **65**, 117 (1944). See also B. Kaufmann, *Phys. Rev.* **76**, 1232 (1949).

6. S. Stein, *Archimedes: What Did He Do Beside Cry Eureka?* (Math. Assn. Am., Washington, D. C., 1999).

7. For X-ray crystal calibration see Yu. V. Shvyd'ko et al., *Phys. Rev. Lett.* **85**, 495 (2000) and references therein.

8. Robert Hooke observed in 1665 that crystals might be packings of identical "globules" of matter. See R. Hooke, *Micrographia* (Science Heritage Ltd., Lincolnwood, IL, 1987).

9. There are many excellent texts on X-ray crystallography. See B. D. Cullity, S. R. Stock, and S. Stock, *Elements of X-Ray Diffraction* (Prentice Hall, New York, 2001) and J. Als-Nielson and D. McMorrow, *Elements of Modern X-Ray Physics* (Wiley, New York, 2001).

10. There is an enormous literature on liquid helium. See, for example, J. F. Allen, *Superfluid Helium* (Academic Press, Burlington, MA, 1966) and J. Wilkes, *The Properties of Liquid and Solid Helium* (Oxford U. Press, London, 1967). The original reference for the discovery of superfluidity in ^{4}He is P. Kapitsa, *Nature* **141**, 79 (1938). For the theory of superfluidity in ^{4}He see I. M. Khalatnikov, *An Introduction to the Theory of Superfluidity* (Benjamin, New York, 1966) and D. Pines and P. Nozières, *The Theory of Quantum Fluids* (Benjamin, New York, 1966).

11. The slowing-down of crystallization in polymers and glasses is one of the things that makes them so useful. But both do crystallize. See J. Schultz, *Polymer Crystallization: The Development of Crystalline Order in Thermoplastic Polymers* (Oxford U. Press, Oxford, 2001) and I. Gutzow, *The Vitreous State: Thermodynamics, Structure, Rheology, and Crystallization* (Springer, Heidelberg, 1995).

12. Protein crystallography is a black art poorly understood by most physicists. See T. M. Bergfors, ed., *Crystallization of Proteins: Techniques, Strategies, and Tips* (International University Line, La Jolla, 1998) and A. McPherson, *Crystallization of Biological Macromolecules* (Cold Spring Harbor Laboratory, Woodbury, NY, 1999).

13. For information on detection of atomic motion by means of inelastic X-ray scattering, see M. Holt et al., *Phys. Rev. Lett.* **83**, 3317 (1999) and references therein.

14. The physics literature on phase transitions is unfortunately technical and opaque. Some key references are H. E. Stanley, *Introduction to Phase Transitions and Critical Phenomena* (Oxford U. Press, London, 1997) and S. Sachdev, *Quantum Phase Transitions* (Cambridge U. Press, London, 2001).

15. The subject of practical metallurgy is vast and complex. See G. E. Dieter, *Mechanical Metallurgy* (McGraw-Hill, New York, 1986).

16. There are many books on the problem of glasses and glass formation. See, for example, E.-J. Donth, *The Glass Transition: Relaxation Dynamics in Liquids and Disordered*

Materials (Springer, Heidelberg, 2001). The classic reference on ordering in disordered media is S. F. Edwards and P. W. Anderson, *J. Phys. F* **5**, 965 (1975). See also M. Mezard, G. Parisi, and M. A. Virasoro, *Spin Glass Theory and Beyond* (World Scientific, Singapore, 1986); and K. Binder and A. P. Young, Rev. *Mod. Phys.* **58**, 801 (1986).

17. The classic text on hydrodynamics is L. D. Landau and E. M. Lifshitz, *Fluid Mechanics* (Addison-Welsey, Reading, Mass., 1959). See also H. Lamb, *Hydrodynamics* (Dover, New York, 1993).

18. The literature on liquid crystals is vast. See P. Yeh and C. Gu, *Optics of Liquid Crystal Displays* (Wiley, New York, 1999). Further information about the nematic phase may be found in P. G. de Gennes, *The Physics of Liquid Crystals* (Oxford U. Press, New York, 1974); and http://www.lassp.cornell.edu/sethna.OrderParameters/Intro.html.

19. The idea that melting of two-dimensional films might be different from conventional melting was first described in J. M Kosterlitz and D. J. Thouless, *J. Phys. C* **6**, 1181 (1973). That the phase thus generated might be distinct from a conventional liquid was later proposed and elaborated on by David Nelson, Bert Halperin, and Peter Young. See D. R. Nelson, *Phys. Rev. B* **18**, 2318 (1978); A. P. Young, ibid. **19**, 1855 (1979); D. R. Nelson and B. I. Halperin, ibid. **21**, 5212 (1980). Recent experimental work on the hexatic phase may be found in R. Radhakrishnan et al., *Phys. Rev. Lett.* **89**, 076101 (2002) and references therein.

20. Experimental observation of a supersolid phase in ^{4}He was recently claimed by E. Kim and M. H. W. Chan, *Nature* **427**, 225 (2004).

21. For a discussion of clouds and cloud formation see B. J. Mason, *The Physics of Clouds* (Clarendon Press, Oxford, 1971).

22. Buckyball diffraction is described in M. Arndt et al., *Nature* **401**, 680 (1999) and B. Brezger et al., *Phys. Rev. Lett.* **88**, 100404 (2002).

23. P. W. Anderson, *Basic Notions in Condensed Matter Physics* (Addison-Wesley, New York, 1984).

24. The subject of quantized vortices in superfluid ^{4}He has a long history. Most recent work has concentrated on the vortex tangle of superfluid turbulence. See M. R. Smith, *Phys. Rev. Lett.* **71**, 2583 (1993) and M. Tsubota, T. Araki, and S. K. Nemirovskii, *Phys. Rev. B* **62**, 11751 (2000).

25. An understanding of dislocations is central to modern metallurgy and crystal growth technology and is thus explained in most modern texts on solid-state physics. See D. Hull and D. J. Bacon, *Introduction to Dislocations* (Butterworth-Heinemann, Burlington, Mass., 2001) and J. Weertman and J. R. Weertman, *Elementary Dislocation Theory* (Oxford U. Press, London, 1992).

26. One of the best books on the standard model is by one of its inventors: G. t'Hooft, *In Search of the Ultimate Building Blocks* (Cambridge U. Press, London, 1996). See also N. Cottingham and D. A. Greenwood, *An Introduction to the Standard Model of Particle Physics* (Cambridge U. Press, London, 1999). An extremely challenging but comprehensive text is by another inventor: S. Weinberg, *Quantum Theory of Fields, Vols. I–III* (Cambridge U. Press, London, 1995).

27. S. Kauffman, *At Home in the Universe: The Search for Laws of Self-Organization and Complexity* (Oxford U. Press, Oxford, 1996).

CHAPTER 5

1. Two of the most popular modern texts on quantum mechanics are R. Shankar, *Principles of Quantum Mechanics* (Plenum, New York, 1994) and C. Cohen-Tannoudji, B. Din, F. Laloe, and B. Dui, *Quantum Mechanics* (Wiley, New York, 1992).

2. This excludes drug-induced ones, which do not count. As a parent, I feel obligated to counter this joke with the frank admission that my wife and I run a zero-tolerance operation and that we even drink very little.

3. Abbott and Costello's "Who's on First" routine was first performed live on radio and subsequently incorporated into the movie *The Naughty Nineties*. It is reproduced at hundreds of sites on the web—far too many to list.

4. The Copenhagen interpretation of quantum mechanics as developed by Bohr, Heisenberg, and Born is a large subdiscipline of the philosophy of science. The best reference is on the web: J. Fain, "The Copenhagen Interpretation of Quantum Mechanics," in *The Stanford Encyclopedia of Philosophy*, E. N. Zalta, ed., http://plato.stanford. edu/archives/sum2002/entries/qm-copenhagen. See also J. Faye, *Neils Bohr: His Heritage and Legacy. An Antirealist View of Quantum Mechanics* (Kluwer, Dordrecht, 1991).

5. There are many reprintings of Berkeley's works. See G. Berkeley and J. Dancy, ed., *A Treatise Concerning the Principles of Human Knowledge* (Oxford U. Press, London, 1998).

6. The original reference for Schrödinger's cat is E. Schrödinger, *Naturewis-senschaften* **23**, 807 (1935), translated by John D. Trimmer in *Proc. Am. Phil. Soc.* **124**, 323, and reprinted as Section I.11 of Part I of *Quantum Theory and Measurement*, J. A. Wheeler and W. H. Zurek, eds. (Princeton University Press, Princeton, 1983). In this article, Schrödinger refers to the cat example as "ridiculous," a fact sometimes overlooked when the story is recounted.

7. It is no longer common for people to know that a Geiger-Müller counter is a detector of ionizing radiation. See G. F. Knoll, *Radiation Detection and Measurement* (Wiley, New York, 2000).

8. Transcripts and recordings of Burns and Allen routines may be found all over the web. See also C. Blythe and S. Sackett, *Say Goodnight Gracie: The Story of Burns and Allen* (E. P. Dutton, New York, 1986).

9. This is actually a gross underestimate. Assuming that every beach is 100 meters wide and 1 meter deep, that there are 100,000 kilometers of beaches in the world, and that each grain of sand has a volume of 1 cubic millimeter, one obtains 10^{19} grains. This is only the number of molecules in a volume of air the size of a sugar cube. Estimates of the number of sand grains on all the beaches in all the world run as high as 10^{22}. This is indeed the number of atoms in a sugar cube, but if one wants to reckon on the number of electrons and count the space dimensions properly, then one needs about ten times this number. See http://www.ccaurora.edu/ast102/notes/notes11.htm and http:// www.tufts.edu/as/physics/courses/physics5/estim_97.html.

10. Entanglement is a timely topic at the moment because of its relevance to quantum computing. See A. D. Aczel, *Entanglement: The Greatest Mystery of Physics* (Four Walls Eight Windows, New York, 2002) and G. J. Milburn and P. Davies, *The Feynman*

Processor: Quantum Entanglement and the Computing Revolution (Perseus Publishing, Cambridge, MA, 1999).

11. The subject of quantum noise generated by amplifiers is an entire subdiscipline of physics. The best reference is unfortunately rather technical: H. A. Haus, *Electromagnetic Noise and Quantum Optical Measurements* (Springer, Heidelberg, 2000). See also Y. Yammamoto and H. Haus, *Rev. Mod. Phys.* **58**, 1001 (1986); and H. A. Haus and J. A. Mullen, *Phys. Rev. A* **128**, 2407 (1962).

12. The bowling ball on the hill is a blue-collar version of the problem of a pencil standing on its tip. The maximum time T that a ball of mass m can remain on a hill of size L is $T = \sqrt{L/(32g)}\ln(8m^2L^3g/\hbar^2)$, where g is the acceleration due to gravity and $\hbar$ is Planck's constant. From http://www.brunswickbowling.com we find the mass of a heavy bowling ball to be 7.3 kilograms. With L taken to be 1 meter, we then get a maximum time T of about 9 seconds.

13. There is a bitter controversy over who invented "the wave." One claim is by Krazy George Henderson, who says on his web site that it took place on 15 October 1981 during an American League playoff series between the Oakland A's and the New York Yankees and was televised. He notes that his authorship was later acknowledged on the air by Howard Cosell in a conversation with Don Meredith. The other is by the University of Washington, which claims that Henderson's wave was not complete, and that the first real wave was invented by Rob Weller on 31 October 1981. See http://www.gameops.com/sro/krazy/home.htm and http://depts.washington.edu/hmb/thehmb/history4.shtml.

14. C. G. Rossetti, *Rossetti: Poems* (Knopf, New York, 1993).

15. See, for example, B. S. DeWitt, H. Everett, and N. Graham, *Many-Worlds Interpretation of Quantum Mechanics* (Princeton University Press, Princeton, 1973).

CHAPTER 6

1. I remember this person as Cokie Roberts but cannot find the reference. See C. Roberts, *We Are Our Mothers' Daughters* (William Morrow, New York, 1998).

2. How this maze of wires functions is explained in J. L. Hennessy, D. A. Paterson, and D. Goldberg, *Computer Architecture: A Quantitative Approach* (Morgan Kaufmann, San Francisco, 2002).

3. An excellent reference on semiconductor function and design is C. T. Sah, *Fundamentals of Solid-State Electronics* (World Scientific, Singapore, 1991).

4. To understand why there is swearing, see B. W. Kernighan and D. M. Ritchie, *The C Programming Language* (Prentice Hall, New York, 1988).

5. There is an enormous, highly politicized literature on software monopoly. Some representative works are K. Aulett, *World War 3.0: Microsoft and Its Enemies* (Random House, New York, 2001); R. B. McKenzie, *Trust on Trial: How the Microsoft Case Is Reframing the Rules of Competition* (Perseus Publishing, Cambridge, Massachusetts, 2000); D. B. Kopel, *Antitrust after Microsoft: The Obsolescence of Antitrust in the Digital Era* (Heartland Institute, Chicago, Illinois, 2001); S. J. Liebowitz and S. E. Margolis, *Winners, Losers, and Microsoft* (Independent Institute, Oakland, CA, 2001). The larger implica-

tions of digital ownership are discussed in L. Lessing, *The Future of Ideas* (Random House, New York, 2001).

6. The heat generation by microchips is also a fundamental design constraint. In the spring of 2004 Intel announced that it was halting research on its latest generations of microprocessor designs (code named Tejas and Jayhawk) because of excessive heat production. See the 17 May 2004 issue of the *International Herald Tribune,* http://www.iht.com/articles/50233.html.

7. There is a large literature on quantum computing, fortunately now growing less rapidly. See G. Johnson, *A Shortcut Through Time: The Path to a Quantum Computer* (Knopf, New York, 2003), R. K. Brylinski and G. Chen, *Mathematics of Quantum Computing* (Chapman and Hall, London, 2002), and D. Bouwmeester, A. Ekert, A. Zeilinger, and A. K. Ekert, *The Physics of Quantum Information: Quantum Cryptography, Quantum Teleportation, Quantum Computation* (Springer, Heidelberg, 2000).

8. B. Schneider, *Applied Cryptography: Protocols, Algorithms, and Source Code in C* (Wiley, New York, 1995).

9. The noise issues in analogue computation are explained in B. H. Vassos and G. W. Ewing, *Analog and Computer Electronics for Scientists* (Wiley, New York, 1993).

10. A classic reference on the physics principles of semiconduction is J. C. Phillips, *Bonds and Bands in Semiconductors* (Academic Press, New York, 1973).

11. Ferdinand Braun also invented the oscilloscope. See F. Kurylo, *Ferdinand Braun, a Life of the Nobel Prizewinner and Inventor of the Cathode-Ray Oscilloscope* (MIT Press, Cambridge, Massachusetts, 1981). See also http://www.fbh-berlin.de/englisch/f_braun.htm.

12. Ballistic motion of electrons and holes in semiconductors is detected directly by cyclotron resonance. See G. Landwehr, *Landau Level Spectroscopy: Part II* (North-Holland, Amsterdam, 1990).

13. For a comprehensive review of noncrystalline electronics see J. Kanicki, *Amorphous and Microcrystalline Semiconductor Devices, Volume II: Materials and Device Physics* (Artech, Norwood, MA, 1992).

14. An especially beautiful example of a hydrogen-like line spectrum in phosphorus-doped silicon may be found in G. A. Thomas et al., *Phys. Rev. B* **23**, 5472 (1981).

15. The original reference for Moore's law is G. E. Moore, *Electronics* **38** (1965). See also J. Fallows, *The Atlantic Monthly* **288**, 44 (2001) and http://www.intel.com/research/silicon/mooreslaw.htm.

16. See, for example, J. D. Lindl, *Inertial Confinement Fusion: The Quest for Ignition and Energy Gain Using Indirect Drive* (Springer, Berlin, 1997).

17. There are many popular books on Taoist philosophy. See A. Huang, *The Complete I Ching: The Definitive Translation by the Taoist Master Alfred Huang* (Inner Traditions Intl. Let., Rochester, Vermont, 1998).

CHAPTER 7

1. Information about the Max Planck Institute in Stuttgart may be obtained from http://www.mpi-stuttgart.mpg.de.

2. This refers to the famous remark by Defense Secretary Rumsfeld at a briefing at the Pentagon Foreign Press Center on 22 January 2003. In response to a question about the lack of enthusiasm in France and Germany for the Iraq invasion, he labeled these countries "Old Europe." See http://www.defenselink.mil/transcripts/2003/t01232003_t0122sdfpc.html.

3. The previous winners since 1945 were Walther Bothe in 1954, Rudolph Mössbauer in 1961, and J. Hans Jensen in 1963. Max Born also won in 1954, but by then had become a British citizen. There have been five German winners since von Klitzing. See http://www.nobel.se.

4. The original paper announcing the discovery of the quantum Hall effect was K. von Klitzing, G. Dorda, and M. Pepper, *Phys. Rev. Lett.* **45**, 494 (1980).

5. The quantized Hall formula is $R = h/ne^2$, where n is an integer.

6. Bell Labs had many fascinating traditions in those days in addition to the tea room. See J. Bernstein, *Three Degrees Above Zero* (Cambridge U. Press, London, 1987).

7. I wrote the first paper linking accuracy of the von Klitzing effect with localization. See R. B. Laughlin, *Phys. Rev. B* **23**, 5632 (1981).

8. The discovery of the fractional quantum Hall effect was announced in D. C. Tsui, H. L. Stormer, and A. C. Gossard, *Phys. Rev. Lett.* **48** 1559 (1982).

9. The qualitative behavior of the von Klitzing effect was anticipated theoretically in T. Ando, *J. Phys. Soc. Japan* **37**, 622 (1974).

10. My original theory paper for the fractional quantum Hall effect is R. B. Laughlin, *Phys. Rev. Lett.* **50**, 1395 (1983).

11. See W. P. Su, J. R. Schriefer, and A. J. Heeger, *Phys. Rev. B.* **42**, 1698 (1979), and references therein.

12. The number of references is unfortunately also cascading. A good representative example is J. P. Eisenstein and H. L. Stormer, *Science* **248**, 1461 (1990).

13. Mainau is a large island on Lake Constance owned by the Count and Countess Bernadotte, patrons of the famous Lindau meetings of Nobel laureates. Its famous gardens are open to the public. See http://www.mainau.de.

CHAPTER 8

1. L. Hoddeson and V. Daitch, *True Genius: The Life and Science of John Bardeen* (Joseph Henry Press, Princeton, NJ, 2002).

2. Frederick Sanger also won the Nobel Prize twice in the same field—first in 1958 for protein structure studies and again in 1980 for recombinant DNA. Linus Pauling won a Nobel Prize in chemistry in 1954 for work on the nature of the chemical bond and again in peace in 1963 for activities against nuclear weapons.

3. My specific sources are J. C. Phillips, C. N. Herring, and T. Geballe.

4. For a brief account of the invention of the transistor see W. F. Brinkman, "The Transistor: 50 Glorious Years and Where We're Going," http://www.lucent.com/minds/transistor/pdf/first50.pdf.

5. See M. Riordan and L. Hoddeson, *Crystal Fire: The Birth of the Information Age* (Norton, New York, 1997); F. M. Wanlass and C. T. Sah, "Nanowatt Logic Using Field-Effect

Metal-Oxide-Semiconductor Transistors," Tech. Dig. IEEE Int. Solid State Circuits Conf., 32–33, 1963.

6. William Shockley had a notorious and colorful career. He emigrated to California, where he seeded the birth of silicon valley. He also became obsessed with the influence of heredity on intelligence. See W. Shockley and R. Pearson, *Shockley on Eugenics and Race: The Application of Science to the Solution of Human Problems* (Scott-Townsend Publishers, Washington, D.C., 1992).

7. I heard this story from Scalapino at a dinner party in 2001.

8. The war-making ability of the executive is a particularly sensitive issue at the moment because of the conflict in Iraq, but a great deal has been said about it previously. See A. M. Schlesinger, Jr., *The Imperial Presidency* (Houghton Mifflin, New York, 1989); and A. Hamilton, J. Madison, and J. Jay, *The Federalist Papers* (Mento, New York, 1961). See also http://www.ciaonet.org/pbei/cato/heq01.pdf.

9. K. Orrison, *Written in Stone: Making Cecil B. DeMille's Epic, The Ten Commandments* (Vestal Press Ltd., Vestal, New York, 1999).

10. A nice account of this story may be found on Bob Schrieffer's web site, http://www.research.fsu.edu/researchr/winter2002/schrieffer.html.

11. I got this idea from a wonderful web site called *Marxist Jeopardy*. See http://www.anzwers.org/free/marx.

12. One quintillion is $10^{18} = 1,000,000,000,000,000,000$.

13. F.-M. A. Voltaire, *Candide or Optimism: A Fresh Translation, Backgrounds, Criticism* (W. W. Norton, New York, 1991).

14. The original reference for the discovery of high-temperature superconductivity is J. G. Bednorz and K. A. Müller, *Z. Phys.* B **64**, 189 (1986).

15. T. Kuntz, "Word for Word—The World's 'Funniest' Jokes: So this German Goes into a Bar with Dr. Watson and a Chicken," *New York Times*, 27 January 2002.

16. This is an excellent example of Nietzsche's aphorism, "The attraction of knowledge would be small if one did not have to overcome so much shame on the way." See F. Nietzsche, *Beyond Good and Evil: Prelude to a Philosophy of the Future*, W. Kaufmann, ed. (Cambridge U. Press, London, 2001).

17. This effect is called White Night in Russia. Its maximum at the solstice is celebrated as a national holiday in Sweden.

CHAPTER 9

1. The direct cost of nuclear weaponry since 1940 has been estimated by a recent Brookings Institution study to be over $5 trillion. See S. I. Schwartz, *Atomic Audit: The Costs and Consequences of U. S. Nuclear Weapons Since 1940* (Brookings Inst. Press, Washington, D.C., 1998). The costs of "supporting" science are more difficult to estimate because of ambiguity over what support means. The FY 2002 Department of Energy budgets were $700 million for particle physics, $300 million for nuclear physics, and $300 million for fusion. See http://www.aip.org/enews/fyi/2001/134.html.

2. The clearest textbook on nuclear physics continues to be E. Segre, *Nuclei and Particles: An Introduction to Nuclear and Subnuclear Physics* (Benjamin Cummings, San Francisco, 1977).

3. The classic reference on the low-energy quantum mechanics of matter is C. Kittel, *Quantum Theory of Solids* (Wiley, New York, 1987). See also J. R. Schrieffer, *Theory of Superconductivity* (Benjamin, New York, 1983).

4. There is an enormous literature on ^{3}He. See D. Vollhardt and P. Wölfle, *The Superfluid Phases of Helium 3* (Taylor and Francis, London, 1990); D. D. Osheroff, *Rev. Mod. Phys.* **69**, 667 (1997); G. E. Volovik, *Exotic Properties of Superfluid ^{3}He* (World, Singapore, 1998). See also http://boojum.hut.fi/research/theory.

5. The liquid phase is well known, but the gas phase has only recently been discovered. It is usually referred to as an atomic "Bose–Einstein condensate." See M. H. Anderson et al., *Science* **269**, 198 (1995).

6. There is an extensive literature on neutron matter and the interior of neutron stars. See J. Saham, *J. de Phys.* **41**, C2–9 (1980) and J. A. Sauls, "Superfluidity in the Interiors of Neutron Stars," in *Tuning Neutron Stars*, H. Ogelman, E. Van den Heuvel, and J. van Paradis, eds. (Kluwer, Dordrecht, 1989), pp. 441–490.

7. See A. D. Kaminker et al., *Astron. Astrophys.* **343**, 1009 (1999).

8. The formulas for the thermal energy per unit volume of light and sound are $u^{light} = ((\pi^2/15)(k_B T)/(\hbar c)^3$ and $u^{sound}/u^{light} = (c/v_t)^3 + 0.5(c/v_l)^3$, where v_t and v_l are the transverse and longitudinal speeds of sound.

9. One of my favorites is the use of a single phonon generated by a spin flip to measure the thickness of a helium film only a few atoms thick. See E. S. Sabisky and C. H. Anderson, *Phys. Rev. A* **7**, 790 (1973). See also D. J. Bishop and J. D. Reppy, *Phys. Rev. Lett.* **40**, 1727 (1978) and references therein.

10. See P. M. Watkins, *Story of the W and Z* (Cambridge U. Press, London, 1986).

11. The equivalence of spontaneous symmetry-breaking in the Higgs mechanism in the physics of superconductors was first pointed out in P. W. Anderson, *Phys. Rev.* **130**, 439 (1963).

12. For a brief mathematical description of plasmons see A. A. Abrikosov, L. P. Gorkov, and I. Dzyaloshinskii, *Methods of Quantum Field Theory in Statistical Physics* (Dover, New York, 1963), p. 195.

13. H. P. J. Wijn, ed. *Landolt-Börnstein, Group III: Crystal and Solid State Physics, Vol 19: Magnetic Properties of Metals, Subvolume d1: Rare Earth Elements, Hydrides and Mutual Alloys* (Springer, Berlin, 1991). There is a huge literature on the magnetic properties of rare earth compounds and alloys. See J. Jensen and A. R. Mackintosh, *Rare Earth Magnetism* (Clarendon Press, Oxford, 1991). The original discovery of spiral antiferromagnetism in elementary holmium is W. C. Koehler et al., *Phys. Rev.* **151**, 414 (1966). More recent investigations have found further mild low-temperature phase transitions associated with commensuration of the spiral with the underlying atomic lattice. See R. A. Cowley and S. Bates, *J. Phys. C* **21**, 4113 (1988) and D. Gibbset et al., *Phys. Rev. Lett.* **55**, 234 (1985).

14. These particular euphemisms are from H. Noel, "The Front Porch—Euphemisms," which appeared in *Senior World Online*, http://www.seniorworld.com/articles/a19991013195512.html.

15. C. W. Kim, *Neutrinos in Physics and Astrophysics* (Harwood Academic, London, 1993).

CHAPTER 10

1. A good discussion of relativity may be found in most elementary college-level physics texts. The original reference is A. Einstein, *Ann. d. Physik* **17**, 891 (1905).

2. A. S. Eddington, *The Mathematical Theory of Relativity* (Cambridge University Press, London, 1965), p. 88.

3. Symmetry is fun to write about, so there are plenty of good books on the subject. A good lay-accessible one is L. M. Lederman and C. T. Hill, *Symmetry and the Beautiful Universe* (Prometheus Books, Amherst, NY, 2004). A good technical text is J. Rosen, *Symmetry Discovered* (Cambridge University Press, London, 1975). See also S. Coleman, *Aspects of Symmetry: Selected Erice Lectures* (Cambridge University Press, London, 1985).

4. R. P, Feynman et al., *Six Not-So-Easy Pieces, Einstein's Relativity, Symmetry, and Space-Time* (Perseus, New York, 1997).

5. The most famous experimental test of general relativity involves small static corrections to Newtonian gravity, notably the bending of light as it grazes the sun and the advance of Mercury's perihelion, first calculated by Einstein himself. A more contemporary test is the gyroscope precession effect being tested by the recently launched Gravity-Probe B experiment. See R. A. Van Patten and C. W. F. Everitt, *Phys. Rev. Lett.* **36**, 629 (1976).

6. The binary pulsar PSR 1913+16, discovered by R. Hulse and J. Taylor, moves in such a tight orbit that the effects of gravitational radiation emission are measurable. The observation won Taylor and Hulse the 1993 Nobel Prize in physics. This pulsar rotates at 17 rotations per second, which corresponds to a 59-millisecond period, and has an orbital period of 7.75 hours. The advance of its perihelion due to emission of gravitational radiation is 4.2 degrees per year. The orbital radius is 3 light seconds, or a million kilometers. See J. H. Taylor, L. A. Fowler, and J. M Weisberg, *Nature* **277**, 437 (1979); J. M Weisberg, J. H. Taylor, and L. A. Fowler, *Scientific American* **245**, 74 (1981).

7. The original mechanical detectors of gravitational waves proved to lack adequate sensitivity, and have since been supplanted in the United States by the Laser Interferometer Gravitational-Wave Observatory project (LIGO), which hopes eventually to directly detect gravitational waves generated by astrophysical sources. See http://www.ligo.caltech.edu.

8. Albert Einstein's first papers were not well understood in part because they were so tightly reasoned. When asked about what he thought of Einstein's ideas, Edison is reported to have said that he did not understand any of them and could see no profit therein. See http://www./patentlessons.com/Warp%20speed.htm.

9. Or Molière's famous example that a sleeping potion works because of its "dormative properties."

10. For an excellent review, see S. Perlmutter, Supernovae, Dark Energy, and the Accelerating Universe, *Physics Today*, April 2003, p. 53. See also S. Perlmutter et al., *Nature* **391**, 51 (1998).

11. This famous story is actually untrue. No such law was ever passed. The story centers on House bill 246 of 1897 introduced by Edwin J. Goodman of Solitude, Indiana. It did not declare π to be 3 but rather one of several values, depending on circumstances. It was passed unanimously by the state Assembly but died in the Senate. See U. Dudley, *Mathematical Cranks* (Math. Assn. Am., Washington, D.C., 1992).

12. For a good overview of supersymmetry see S. Weinberg, *Quantum Theory of Fields, Vol. 3: Supersymmetry* (Cambridge, University Press, London, 2000).

CHAPTER 11

1. A good book on fractals is G. W. Flake, *The Computational Beauty of Nature: Computer Explorations of Fractals, Chaos, Complex Systems and Adaptation* (MIT Press, Cambridge, 1998). Sell also B. B. Mandelbrot, *The Fractal Geometry of Nature* (W. H. Freeman, New York, 1982). There is a lot of fractal art on the internet. See, for example, http://pages.globetrotter.net/mdessureault/vent.htm and http://www.fractalus.com/galleries/home.

2. Stephen Wolfram felt so strongly that there *was* a new kind of science that he published his own book: S. Wolfram, *A New Kind of Science* (Wolfram Research, Champaign, IL, 2002). See also S. Wolfram, *Nature* **311**, 419 (1984).

3. The fundamental idea behind fractal structures is self-similarity. See M. Ausloos and D. H. Berman, *Proc. Roy. Soc. [London] A* **400**, 331 (1985). The best references for fractal mountains are on the internet. See http://www.skytopia.com/gallery/mountains/mountains.html. A good explanation of how fractal mountains are made is found at http://www.mactech.com/articles/mactech/mactech/Vol.07/07.05/FractalMountains/. Fractal coastlines are explained at http://polymer.bu.edu/ogaf/html/cp2.htm.

4. The best overview of diffusion-limited aggregation is T. C. Halsey, *Physics Today* **53**, 36 (November 2000). The original reference is T. A. Witten, Jr., and L. M. Sander, *Phys. Rev. Lett.* **47**, 1400 (1981); See also P. Meakin, *Phys. Rev. A* **27**, 1495 (1983).

5. See M. Gardner, *Wheels, Life, and Other Mathematical Amusements* (W. H. Freeman, New York, 1983). The original reference is M. Gardner, *Scientific American* **223**, 120 (October 1970). See also E. R. Berlekamp, J. H. Conway, and R. K. Gray, *Winning Ways for Your Mathematical Plays, II: Games in Particular* (Academic Press, Burlington, MA, 1982); and J. Conway, *On Numbers and Games* (Academic Press, Burlington, MA, 1976). There is an immense amount of material on Conway's *Life* on the internet. A good place to start is http://www.radicaleye.com/lifepage. See also http://www.argentum.freeserve.co.uk/lex.htm.

6. Nanotubes are a subject of intense interest in the research community at the moment. See M. S. Dresselhaus, G. Dresselhaus, and P.C. Eklund, *The Science of Fullerenes and Carbon Nanotubes* (Academic Press, Burlington, MA, 1996). The original discovery paper is S. Iijima, *Nature* **354**, 56 (1991).

7. I am not making this up. See Mike Martin's article at *Wireless NewsFactor*, http://www.wirelessnewsfactor.com/perl/story/20867.htm. Other proposed uses for nanotubes include field emitters for displays, conductive plastics, energy storage (bat-

teries), molecular electronics, thermal materials, structural composites, catalyst supports, and sensors.

8. Nanopeapods are nanotubes with buckyballs stuffed inside. See B. W. Smith and D. E. Luzzi, *Chem. Phys. Lett.* **321**, 169 (2000).

9. A representative publication is M. Bockrath et al., *Phys. Rev. B* **61**, 10606 (2000). See also http://smalley.rice.edu.

10. The most comprehensive reference is the web page of the Alivisatos group at UC Berkeley: http://www.cchem.merkeley.edu/~pagrp/overview.html. See also B.O. Dabbousi et al., *J. Phys. Chem. B* **101**, 9463 (1997).

11. This is the procedure originally used to make porous silicon, the source of silicon nanocrystals. See L. T. Canham, *Appl. Phys. Lett.* **57**, 1046 (1990).

12. That quantum-mechanical electrons should crystallize if sufficiently cold and dilute was first realized by the theoretician E. P. Wigner. The observation of Wigner crystallization was through electrons sprayed on the surface of liquid helium. See C. C. Grimes and G. Adams, *Phys. Rev. Lett.* **42**, 795 (1970).

CHAPTER 12

1. David Pines and I coined the term "protection" as a lay-accessible synonym for the technical (and thus confusing) physics term "attractive fixed point of the renormalization group." See R. B. Laughlin and D. Pines, *Proc. Natl. Acad. Sci.* **97**, 28 (2000).

2. These relationships are explained succinctly in P. W. Anderson, *Concepts in Solids* (World Scientific, Singapore, 1998).

3. Harlan Ellison's novella *A Boy and His Dog* was made into a low-budget movie starring Don Johnson in 1974. The story originally appeared in H. Ellison, *The Beast That Shouted Love and the Heart of the World* (Avon Books, New York, 1969).

4. There is an enormous literature on the subject of scale invariance and renormalizability in phase transitions. The text I usually recommend is by one of its discoverers: L. P. Kadanoff, *Statistical Physics: Statics, Dynamics and Renormalization in Statistical Physics* (Cambridge University Press, London, 1996). Note that it is generally understood that the *quantum* (i.e., zero-temperature) versions of these phenomena are qualitatively similar to the "statistical" (i.e., finite-temperature) ones. See S. Sachdev, *Quantum Phase Transitions* (Cambridge University Press, London, 2000).

5. J. C. Collins et al., *Renormalization* (Cambridge University Press, London, 1984). See also C. Itzykson et al., *Statistical Field Theory: Volume 1, From Brownian Motion to Renormalization and Lattice Gauge Theory* (Cambridge University Press, London, 1989) and J. Cardy et al., *Scaling and Renormalization in Statistical Physics* (Cambridge University Press, London, 1996).

6. The classic occurrence of critical opalescence is in hot compressed gases. For example, light scattering from carbon dioxide at its critical point is reported in J. A. White and B. S. Maccabee, *Phys. Rev. Lett.* **26**, 1468 (1971). More accessible examples occur in chemical systems: P. A. Egelstaff and G. D. Wingnall, *J. Phys. C* **3**, 1673 (1973); J. S. Huang and M. W. Kim, *Phys. Rev. Lett.* **47**, 1462 (1981); C. Herkt-Maetzky

and J. Schelton, *Phys. Rev. Lett.* **51**, 896 (1983); G. Dietler and D. S. Cannell, *Phys. Rev. Lett.* **60**, 1852 (1988).

7. The sketch "Hunting the Deceitful Turkey" first appeared in *The Mysterious Stranger*. It is reprinted in *Mark Twain: Collected Tales, Sketches, Speeches and Essays*, L. J. Budd, ed. (Library of America, 1992) and is also available on the internet at http://www.gutenberg.org/etext/3186.

8. I am infringing here on Karl Popper's philosophy of scientific epistemology, which is the subject of endless discussion among academics. I will give here only the original source: Popper's book *Logik der Forschung*, reprinted as K. Popper, *The Logic of Scientific Discovery* (Routledge, NY, 2002).

9. The literature on the correlated-electron problem on the internet is confusing and much too extensive to reference. For a sensible overview of the field I recommend Z. Wang et al., *Strongly Correlated Electronic Materials* (Westview Press, Boulder, CO, 1994).

10. The original solution of the silicon 7×7 problem is K. Takayanagi, Y. Tanishiro, S. Takahashi, and M. Takahashi, *Surf. Sci.* **164**, 367 (1985). A relevant theoretical paper from the time is I. Stich et al., *Phys. Rev. Lett.* **68**, 1351 (1992).

11. An up-to-date discussion of cosmological issues, including the relevance of vacuum renormalizability, may be found in G. W Gibbons et al., eds., *The Future of Theoretical Physics and Cosmology: A Celebration of Stephen Hawking's 60th Birthday* (Cambridge University Press, London, 2003).

CHAPTER 13

1. For an introduction to this subject, see M. Schena, *Microarray Analysis* (Wiley-Liss, New York, 2002).

2. Flash memory has become wildly popular recently in the form of USB memory sticks. See P. Cappellelti et al., *Flash Memories* (Kluwer, Amsterdam, 1999).

3. There is an enormous literature on protease inhibitors. See, for example, R. C. Ogden and C. W. Flexner, eds., *Protease Inhibitors in AIDS Therapy* (Marcel Dekker, New York, 2001).

4. Stem cell research is highly controversial and thus in the news at the moment. A comprehensive survey from the perspective of the National Institutes of Health is available from their web site: *Stem Cells: Scientific Progress and Future Directions*, http://www.nih.gov/news/stemcell/scireport.htm.

5. This is the famous golden rice. See M. L. Guerinot, *Science* **287**, 241 (2000); X. Ye et al., *Science* **287**, 241 (2000). There is political opposition to this particular biotech product. See http://www.biotech-info.net/golden.html.

6. M. W. Shelley, *Frankenstein, or the Modern Prometheus* (Palgrave Macmillan, New York, 2000). A great deal has been written about this astonishing novel. See M. Spark, *Mary Shelley* (Meridian, New York, 1988); http://www.kimwoodbridge.com/maryshel/essays.shtml; http://home–1.worldonline.nl/~hamberg.

7. R. J. Jackson et al. *J. Virol.* **75**, 1205 (2001). The accidental creation of a deadly variant of the mousepox virus by recombinant methods has sparked a heated public

debate about the dangers of biotechnology and the need for strong classification rules. See J. Cohen, "Designer Bugs," *Atlantic*, July–August 2002, p. 113. The mousepox story acquired a horrifying twist recently when a team under Professor Mark Buller at St. Louis University repeated the experiment. See W. J. Broad, "Bioterror Researchers Build a More Lethal Mousepox," *New York Times*, 1 November 2003.

8. See E. Teller and J. Shoolery, *Memoirs: A Twentieth-Century Journal of Science and Politics* (Perseus Press, Cambridge, Massachusetts, 2002).

9. I obtained the billion-dollar estimate from the annual report of Affymetrix Corporation, the major worldwide supplier of cDNA arrays, available at http://biz.yahoo.com/e/010515/affx.htm. It reports recent profits from sales to be about $200 million per year. I assumed this number also to be net sales in chips, since arrays are its largest-volume sales item and are almost pure profit. The market price of these varies, but is reported to be about $1,000. (See http://www.research.bidmc.harvard.edu/corelabs/genomic/default.asp.) That means sales of about 200,000 gene chips per year, and thus 200,000 experiments. Factoring in labor, laboratory costs, and overhead, I estimate each of these experiments to cost $50,000. As an independent check I note that that the NIH budget for FY 2001 was $19 billion, 81% of which was extramural research. That would make array work about 7% of the total extramural expenditure, which is a reasonable estimate.

10. The famous dog that did not bark appears in Arthur Conan Doyle's story "Silver Blaze." See A. C. Doyle, *Complete Sherlock Holmes* (Doubleday, New York, 2002).

11. An autopilot is a specific example of feedback control. See S. Skosestad, *Multivariate Feedback Control* (Wiley, New York, 2005).

12. For an explanation of amplifiers see S. Franco, *Design with Operational Amplifiers and Analog Integrated Circuits* (McGraw Hill, New York, 1997).

13. See, for example, A. Fersht, *Structure and Mechanism in Protein Science: A Guide to Enzyme Catalysis and Protein Folding* (W. H. Freeman, New York, 1999) and A. M. Lesk, *Introduction to Protein Architecture: The Structural Biology of Proteins* (Oxford U. Press, London, 2001).

14. The original idea of this motor was published by Paul Boyer in 1964. One of its parts was subsequently crystallized by John Walker. The two of them shared the 1997 Nobel Prize in chemistry for working out the function of this enzyme. See P. D. Boyer, *Angew. Chem. Int. Ed.* **37**, 2296 (1998); J. E. Walker, ibid., 2308. A key experiment confirming its mechanical nature was done by Masasuke Yoshida, who attached an actin filament to the rotor and observed it revolving under a microscope. See H. Noji, R. Yasuda, M. Yoshida, and K. Kinosita, Jr., *Nature* **386**, 299 (1997); and http://www.res.titech.ac.jp. See also H. Wang and G. Oster, *Nature* **396**, 279 (1998); and H. Seelert et al., ibid. **405**, 418 (2000).

15. There is a truly stupendous literature on motor proteins. I recommend starting with Eckhard Jankowsky's excellent web site http://www.helicase.net/dexhd/motor.htm. The pioneering work on actin-myosin was done by Jim Spudich and is reviewed in J. A. Spudich, *Nature* **372**, 515 (1994). The kinesin literature is surveyed at http://www.imb-jena.de/~kboehm/Kinesin.html. See also K. Kawaguchi and S. Ishiwata, *Science* **291**, 667 (2001) and references therein.

16. See H. Salman, Y. Soen, and E. Braun, *Phys. Rev. Lett.* 77, 4458 (1996), and references therein.

17. W. E. Stegner, *Beyond the Hundredth Meridian: John Wesley Powell and the Second Opening of the West* (Penguin, New York, 1992).

CHAPTER 14

1. A nice list of rock stars who suffered premature death may be found at http://elvispelvis.com/fullerup.htm. Other sites on this topic include http://www.av1611.org/rockdead.html and http://www.wikipedia.org/wiki/List_of_artists_who_died_of_drug-related_causes.

2. The Los Alamos bulletin board was originally given the URL http://xxx.lanl.gov in order that it would come up frequently in searches. It has subsequently moved to Cornell and now has the URL http://arxiv.org. Biographical information about Professor Ginsparg may be found at http://www.physics.cornell.edu/profpages/Ginsparg.htm.

3. The Gordon and Betty Moore Foundation recently gave a grant of $9 million to the Public Library of Science for the creation of two refereed electronic journals, *PLoS Biology* and *PLoS Medicine*. See http://www.bio-itworld.com/archive/021003/firstbase.html.

4. Bolt, Beranek, and Newman (BBN) is the company in Cambridge, Massachusetts, that was awarded the original contract to build the Darpanet. See http://www.bbn.com.

5. For an introduction to fusion engineering see A. A. Harms et al., *Principles of Fusion Energy* (Wiley, New York, 2000).

6. Cold fusion was announced by Stanley Pons and Martin Fleischmann in a news conference in March 1989. They subsequently obtained $5 million research money from the Utah state legislature, $500 thousand of which came from an anonymous private donor. It later came to light that this increment had actually come from the university's own research funds. Earlier in his career, Fleischmann had discovered something called the surface-enhanced Raman effect, which is both legitimate and technologically useful. See M. Fleischmann, P. J. Hendra, and A. J. McQuillan, *Chem. Phys. Lett.* **26**, 163 (1974).

7. See J. R. Huizenga, *Cold Fusion: The Scientific Fiasco of the Century* (Oxford University Press, London, 1994).

8. See http://www.sciencefriday.com/pages/1997/Apr/hour_1_041197.htm. The broadcast of NPR's *Science Friday* with Ira Flatow in question took place on 11 April 1997.

9. See http://www.infinite-energy.com.

10. W. J. Broad, *Star Warriors* (Simon and Schuster, New York, 1986); ibid., *Teller's War: The Secret Story Behind the Star Wars Deception* (Simon and Schuster, New York, 1992). See also C. E. Bennett's article "The Rush to Deploy SDI," in the April 1988 issue of the *Atlantic Monthly*, http://www.theatlantic.com/issues/88apr/bennett_p2.htm.

11. See http://www.nrdc.org/nuclear/nif2/findings.asp. The National Ignition Facility at Livermore is strongly opposed by the Natural Resources Defense Council.

12. A great deal has been written about the Schön affair. See K. Chang, "Panel Says Bell Labs Scientist Faked Discoveries," New York Times 26 September 2002. The official

statement from Lucent accepting culpability may be found at http://www.lucent. com/news_events/researchreview.html. See also R. B. Laughlin, *Physics Today*, December 2002, p. 10, and references therein.

13. See Ian Hoffman's article in the *Oakland Tribune* of 10 September 2002, reproduced at http://www.highfrontier.org/OaklandTribune.9–10–02.htm. See also http://www.periscope1.com/demo/weapons/misrock/antiball/w0003565.html. Brilliant Pebbles was conceived by Lowell Wood and Gregory Canavan in 1986.

14. A classic text on shock waves is Ya. B. Zeldovich, *Physics of Shock Waves and High Temperature Hydrodynamic Phenomena* (Academic Press, Burlington, MA, 1967). See also Ya. B. Zeldovich et al., *Stars and Relativity* (Dover, Mineola, NY, 1997).

15. The best rocket book is still R. H. Goddard, *Rockets* (Dover, Mineola, NY, 2002). See also G. P. Sutton and O. Biblarz, *Rocket Propulsion Elements* (Interscience, New York, 2000).

16. The original text of Robin Erbacher's notorious skit may be found at http://www.stanford.edu/dept/physics/Lighter_Side/Skit.

17. R. Bradbury, *The Martian Chronicles* (William Morrow, New York, 1997).

CHAPTER 15

1. Professor Gumbrecht has very diverse interests, which might by summarized as cautious Epicureanism. He is quoted on the web as wishing to maximize elementary pleasures, such as eating good food and watching sports, to keep alive the sublime complexity of university life, and to maintain the university as a haven for riskful thinking. See H. U. Gumbrecht, *The Powers of Philology: Dynamics of Textual Scholarship* (U. of Illinois Press, Champaign, IL, 2002); T. Lenoir and H. U. Gumbrecht, *Inscribing Science: Scientific Texts and the Materiality of Communication* (Stanford U. Press, Stanford, 1998); H. U. Gumbrecht, *In 1926: Living on the Edge of Time* (Harvard U. Press, Cambridge, 1997). See also http://www.stanford.edu/dept/news/report/news/november29/gumbrecht–1129.html.

2. Professor Linde is especially well known for his contributions to inflationary cosmology, for which he won the Dirac Medal, along with Alan Guth and Paul Steinhardt, in 2002. See A. D. Linde, *Inflation and Quantum Cosmology* (Academic Press, Burlington, MA, 1990).

3. Professor Mitchell is particularly interested in biological self-organization, for example in colonies of social insects. See S. D. Mitchell, *Biological Complexity and Integrative Pluralism* (Cambridge U. Press, Cambridge, 2003).

4. Professor Bravman's interests include electromigration, fatigue in MEMS applications, metal-oxide dielectrics for nanometer-scale transistors, and the mechanical behavior of thin-film interfaces for packaging materials. He is also presently vice provost for undergraduate education at Stanford.

5. Professor Pickstock has written extensively on what she understands to be a need for Western society to be rooted in worship. She also challenges the right of a highly trained scientific elite to define reality. See C. Pickstock, *After Writing: On the Liturgical*

Consummation of Philosophy (Blackwell, Oxford, 1997); and G. Ward, J. Milbank, and C. Pickstock, eds. *Radical Orthodoxy: A New Theology* (Routledge, London, 1999).

6. Professor Djerassi is a two-career scholar, having first invented the birth control pill and then advanced to writing novels and plays. See C. Djerassi, *This Man's Pill: Reflections on the 50th Birthday of the Pill* (Oxford U. Press, London, 2001); C. Djerassi, *The Pill, Pygmy Chimps, and Degas' Horse: The Remarkable Autobiography of the Award-Winning Scientist Who Synthesized the Birth Control Pill* (Basic Books, New York, 1998); C. Djerassi, *Oxygen* (Wiley, New York, 2001).

7. Professor Seel is concerned with how one's philosophical outlook changes what one perceives and, ultimately, the course of one's life. See M. Seel, *Ästhetik des Erscheinens* (Hansen, München, 2000); and M. Seel, *Sich bestimmen lassen: Studien zur theoretische und praktischen Philosophie* (Suhrkamp, Frankfurt, 2002).

8. Professor Schmandt-Besserat has proposed a highly plausible theory that cuneiform writing evolved out of counting conventions required for commerce. See D. Schmandt-Besserat, *How Writing Came About* (U. of Texas Press, Austin, 1996).

9. Professor Ford's chief interests are antidiscrimination and property law. See R. T. Ford, *Racial Culture: A Critique* (Princeton University Press, Princeton, NJ, 2004).

10. Professor Winograd is one of the pioneers of computer intelligence. Two of his students started the internet search company Google. See T. Winograd and F. Flores, *Understanding Computers and Cognition: A New Foundation for Design* (Addison-Wesley, Boston, 1987).

11. Professor Kablitz's main interest is philology, notably the relationship of Dante's *Divina Commedia* to other important works in Western literature, in particular Virgil's *Aeneid* and the Bible. See A. Kablitz and G. Neumann, *Mimesis und Simulation* (Rombach, Freiburg, 1998); A. Kablitz and H. Pfeiffer *Interpretation und Lektüre* (Rombach, Freiburg, 2001).

12. W. Godzich, *The Culture of Literacy* (Harvard U. Press, Cambridge, 1994); W. Godzich and J. Kittay *The Emergence of Prose: An Essay in Prosaics* (U. of Minnesota Press, Minneapolis, 1987).

13. This famous statement appears in Justice Potter Stewart's concurring opinion in the 1964 Supreme Court case Jacobellis versus Ohio. It involved a theater that showed the French movie *The Lovers* containing a brief sex scene with Jean Moreau. Justice Stewart concurred with overturning the decision on the grounds that the film was not hardcore pornography.

14. Some of these statements are taken verbatim from reports generated by committees at the Workshop on Emergence, held at the Stanford Center for the Humanities in August 2002.

15. For a good review of social insect behavior see B. Holldobler and E. O. Wilson, *The Ants* (Bellknap Press, Cambridge, 1990).

16. For a review of theories of the origin of writing see P. T. Daniels and W. Bright, eds., *The World's Writing Systems* (Oxford U. Press, New York, 1996). There is also a terrific web site on this subject maintained by software engineer L. K. Lo. See http://www.ancientscripts.com.

CHAPTER 16

1. Although it first appeared in English in 1635, "lessness" is effectively a neologism. See D. Coupland, *Generation X: Tales for an Accelerated Culture* (St. Martin's Press, New York, 1992).

2. D. Adams, *The Hitchhiker's Guide to the Galaxy* (Ballantine Books, New York, 1995). This book was originally published in 1975 and made into a television series by the BBC.

3. R. Graves, *The Greek Myths, Vol. I* (Penguin Books, Baltimore, MD, 1961), p. 31.

4. The Magnificent Seven are T. Ando, Hiroshi Eisaki, Atsushi Fujimori, Naoto Nagaosa, Tajima, Yoshi Tokura, and Shen-ichi Uchida. It is actually eight if one counts Sadamichi Maekawa, who is a past member.

INDEX